AS/A-LEVEL YEARS 1 AN[D]

STUDENT GUIDE

EDEXCEL

Chemistry

Practical assessment

David Scott

HODDER
EDUCATION
AN HACHETTE UK COMPANY

Hodder Education, an Hachette UK company, Blenheim Court, George Street, Banbury, Oxfordshire OX16 5BH

Orders

Bookpoint Ltd, 130 Park Drive, Milton Park, Abingdon, Oxfordshire OX14 4SE

tel: 01235 827827

fax: 01235 400401

e-mail: education@bookpoint.co.uk

Lines are open 9.00 a.m.–5.00 p.m., Monday to Saturday, with a 24-hour message answering service. You can also order through the Hodder Education website: www.hoddereducation.co.uk

ISBN 978-1-4718-8567-9

First printed 2017

Impression number 5 4 3 2

Year 2020 2019 2018 2017

This guide has been written specifically to support students preparing for the Edexcel AS and A-level Chemistry examinations. The content has been neither approved nor endorsed by Edexcel and remains the sole responsibility of the author.

Cover photo: Ryan McVay/Photodisc/Getty Images/Professional Science

Typeset by Integra Software Services Pvt. Ltd, Pondicherry, India

Printed in Dubai

Hachette UK's policy is to use papers that are natural, renewable and recyclable products and made from wood grown in sustainable forests. The logging and manufacturing processes are expected to conform to the environmental regulations of the country of origin.

Contents

About this book 4

Core Practicals

1 Measure the molar volume of a gas 5

2 Use a standard solution to find the concentration
of a solution of sodium hydroxide 9

3 Find the concentration of a solution of hydrochloric acid . . . 13

4 Investigation of the rates of hydrolysis of halogenoalkanes . . . 15

5 Investigation of the oxidation of ethanol 19

6 Chlorination of 2-methylpropan-2-ol using concentrated
hydrochloric acid 23

7 Analysis of some inorganic and organic unknowns 26

8 To determine the enthalpy change of a reaction
using Hess's law 29

9 Finding the K_a value for a weak acid 32

10 Investigating some electrochemical cells 35

11 Redox titration 38

12 The preparation of a transition metal complex 41

13a Following the rate of the iodine–propanone
reaction by a titrimetric method 44

13b Investigating a 'clock reaction' to determine a
rate equation 49

14 Finding the activation energy of a reaction 53

15 Analysis of some inorganic and organic unknowns 57

16 The preparation of aspirin 59

Questions & Answers

Practical exam-style questions 64

Index 95

■About this book

The purpose of this guide is to help you to prepare for practical-based questions that you will encounter in papers 1 and 2 of the Edexcel AS Chemistry qualification and papers 1, 2 and 3 in the full Edexcel A-level Chemistry qualification.

All papers in the A-level examination will examine 'Working as a chemist'. This means students:

- working scientifically, developing competence in manipulating quantities and their units, including making estimates
- experiencing a wide variety of practical work, developing practical and investigative skills by planning, carrying out and evaluating experiments, and becoming knowledgeable about the ways in which scientific ideas are used
- developing the ability to communicate their knowledge and understanding of chemistry
- acquiring these skills through examples and applications from the entire course

In particular, paper 3 covers the general and practical principles of chemistry. It is of 2 hours and 30 minutes duration and is worth 120 marks. This paper may draw on any of the topics in this specification and includes:

- synoptic questions that may draw on two or more different topics
- questions that assess conceptual and theoretical understanding of experimental methods (indirect practical skills), which will draw on students' experiences of the core practicals

In the AS examination there are just two papers and questions based on practical work may be set in both papers.

During your AS/Year 1 chemistry course you will tackle eight core practical investigations and a further eight practical investigations if you are preparing for the full A-level qualification. You will need to keep a record of your observations and inferences as you progress, and take time to reflect on how the procedures help to illustrate the theoretical ideas you study. If you simply follow practical instructions uncritically, questions on all three papers that require you to have a full appreciation of practical chemistry will catch you out. If, however, you think carefully about why you are carrying out certain procedures and are aware of their possible limitations, you will be in a better position to answer these questions and gain full credit.

This guide has two sections:

- The first section takes you through the eight AS/Year 1 core practical and the eight A-level/Year 2 core practical investigations, as detailed in the Edexcel (8CH0/9CH0) specification. Each practical is considered in context and links are made to the theoretical aspects of the course. The questions posed in each core practical will test your full understanding of the procedure and how improvements might be made. Throughout, issues relating to precision, uncertainty and handling of data appropriately are considered as they are encountered.
- The Questions & Answers section consists of exam-style questions of the type you can expect to encounter in papers 1 and 2 at AS, and in papers 1, 2 and 3 at A-level. Each question includes commentary on how the question should be tackled and how marks are awarded, and is accompanied with a sample answer.

Specification details and supplementary information can be found at
http://qualifications.pearson.com/en/qualifications/edexcel-a-levels/chemistry-2015.html

Core Practicals

■ Core practical 1

Measure the molar volume of a gas

In this practical, known quantities of reactants are combined to produce a gas, which is collected over water. By knowing the relationship between the number of moles of reactants and products as shown in the balanced equation, it is possible to estimate a value for the volume of 1 mole of a gas. Make sure that you are comfortable converting masses of solids into moles and volumes of gases into moles.

For solids (and liquids):

$$\text{number of moles} = \frac{\text{mass in grams}}{\text{mass of 1 mole of substance}}$$

For gases:

$$\text{number of moles} = \frac{\text{volume of gas in dm}^3}{24 \text{ (at 298 K and 101 kPa)}}$$

The procedure

1 Place $30\,\text{cm}^3$ of $1\,\text{mol dm}^{-3}$ ethanoic acid in a boiling tube.
2 Set up the apparatus as shown in Figure 1.

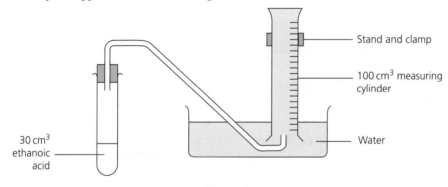

Figure 1

3 Place approximately $0.05\,\text{g}$ of calcium carbonate in a test tube and weigh the test tube and its contents accurately.
4 Removing the bung from the boiling tube, add the contents of the test tube and quickly replace the bung.
5 As the reaction progresses, the gas is collected over water.
6 Once the reaction is complete, measure the volume of gas collected in the measuring cylinder.
7 Reweigh the test tube that contained the calcium carbonate.
8 Repeat the experiment six more times, increasing the mass of the calcium carbonate by about $0.05\,\text{g}$ each time, up to a maximum of $0.40\,\text{g}$.

Core Practicals

The reaction is represented in the following equation:

$$CaCO_3(s) + 2CH_3COOH(aq) \rightarrow (CH_3COO^-)_2Ca^{2+}(aq) + CO_2(g) + H_2O(l)$$

Introductory questions: part 1

a Identify the acid and the base in this reaction and explain why the base cannot be classified as an alkali.

b Re-write the equation above in the form of an ionic equation.

c Calculate the number of moles in 0.36 g of calcium carbonate. Give your answer to the appropriate number of significant figures.

d Calculate the number of moles of ethanoic acid in 30 cm^3 of 1 mol dm^{-3} solution.

e Why is it not important to have the concentration of the acid measured to any greater accuracy than 1 mol dm^{-3}?

f Describe what you would *see* as the reaction proceeds.

g How do you know when the reaction is complete?

Theoretical background

Before we consider this reaction in detail, let us reflect on the concept of the **molar gas volume**. Why is it reasonable to say that 1 mole of *any* gas will occupy the same volume under the same conditions of temperature and pressure?

Introductory questions: part 2

Let us start by considering two gaseous molecules under conditions of 101 kPa and 298 K.

a Work out the M_r of the molecules H_2 and SO_2.

b What assumptions are needed to explain why these two gases occupy the *same* volume at the same temperature and pressure?

c Draw a set of axes with volume of 1 mole of gas on the y-axis and temperature in K on the x-axis and plot the two points for the molar gas volumes given in the margin tip.

d Using the relationship $PV = nRT$, calculate the volume of 1 mole of a gas at 150 K. ($R = 8.31$ J K^{-1} mol^{-1})

e **Extrapolate** the graph you have drawn in part **c** to get a volume for the gas at 150 K. How closely does your value agree with that calculated in part **d**?

Some sample results from CP1

Table 1

	Experiment number					
	1	2	3	4	5	6
Initial mass of test tube and calcium carbonate/g	19.17	19.13	19.04	19.22	19.18	19.42
Final mass of test tube and calcium carbonate/g	19.12	19.04	18.89	19.01	18.90	19.04
Mass of calcium carbonate added/g	0.05	0.09	0.15	0.21	0.28	0.38
Volume of gas collected/cm^3	4	8	15	24	38	75

Practical tip

Remember that an acid is an H$^+$ ion (sometimes referred to as a proton) donor.

Practical tip

Remember that the number of significant figures that you can give your answer is determined by the measurement made to the *least* number of significant figures.

Practical tip

Remember that at 101 kPa and 298 K the volume of 1 mole of any gas can be assumed to be 24 dm^3 and at 100 kPa and 273 K the volume of 1 mole of gas is 22.4 dm^3.

Extrapolate in this case means continuing the line of the graph such that any trend is continued.

Questions on practical data

a Plot a graph of mass of calcium carbonate added (x-axis) against volume of gas collected (y-axis).

b Connect the points using a line of best fit.

c What are the problems in extending this graph to estimate a value for the molar gas volume?

d Use the results above to calculate an *average* value for the molar gas volume. What do you notice?

e Now use the results from experiment 6 only to calculate a value for the molar gas volume.

f Compare your results from parts **d** and **e** and your expected result. What do you notice?

How might this experiment be improved?

In the experiment above, the gas was collected over water. Can you think why this method might be less suitable for some gases than others?

Consider the data in Table 2.

g Referring to Table 2, outline another method for measuring the molar gas volume. You should detail the method and the reactants used and indicate why your method is an improvement on the one outlined above.

Table 2

Gas	Solubility/g per 100g water at 298K and 101KPa
CO_2	0.17
O_2	0.0043
H_2	0.00016

Answers

Introductory questions: part 1

a Ethanoic acid is the acid and calcium carbonate is the base. An alkali is a soluble base and calcium carbonate is insoluble.

b $CaCO_3(s) + 2H^+(aq) \rightarrow Ca^{2+}(aq) + H_2O(l) + CO_2(g)$

c $0.36/M_r$ of $CaCO_3 = 0.36/100.1 = 3.60 \times 10^{-3}$. 2 s.f. is the maximum number of s.f. as the mass is given to only 2.s.f.

d $C = n/V$, so $1 = n/0.03 = 0.03$ moles of acid.

e As the acid is in excess, the concentration of the acid to 1 s.f. is acceptable.

f There would be a slight effervescence.

g The bubbling would stop and the remaining products would be a colourless solution.

Introductory questions: part 2

a M_r of $H_2 = 2$; M_r of $SO_2 = 64.1$

b The assumption is that the actual volume taken up by the molecules themselves is negligible. In other words, the distance between the molecules in the gaseous state is so great that the molecules themselves can be considered as points in space.

c

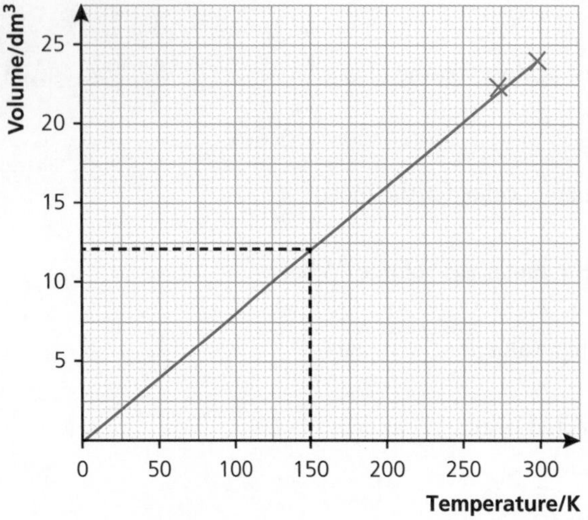

d Using $PV = nRT$, $101 \times V = 1 \times 8.31 \times 150$, $V = 12.3\,dm^3$

e Extrapolation gives 12–$12.5\,dm^3$.

Questions on practical data

a, b

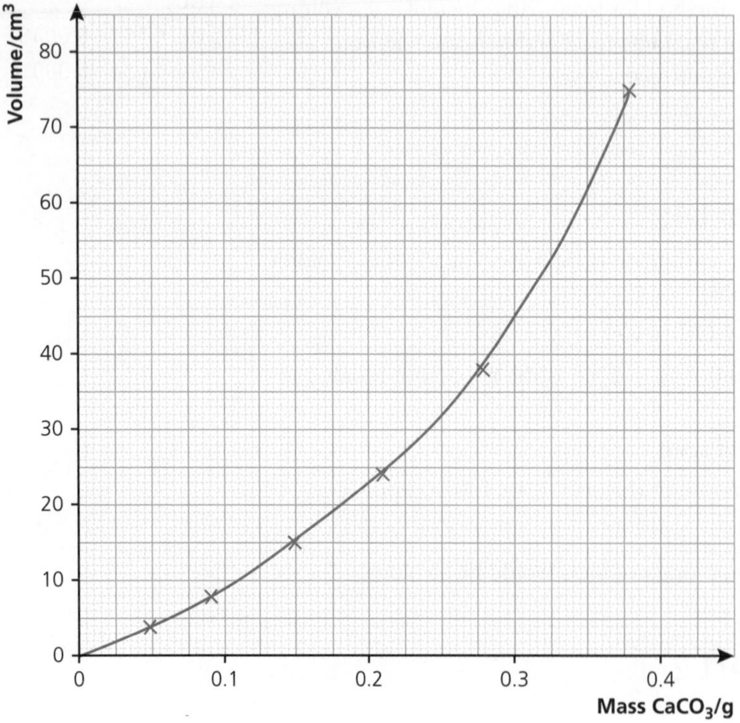

c   It is more difficult to extrapolate, as the relationship is not linear.

d   You can calculate the gas volume but the value is (a) low and (b) depends on which experiment is used. You could use the total volume of gas ($164\,cm^3$) and total mass $1.16\,g$. Then use ratios: $1.16/164 = 100.1/V$, where V = volume of 1 mole = $14.2\,dm^3$.

e   Using experiment 6 only: $(100.1 \times 75)/0.38 = 19.8\,dm^3$.

f   The expected value is $24\,dm^3$, so the value from experiment 6 is closer to the expected result.

How might this experiment be improved?

The main point to notice here is that CP1 gives inconsistent results that are possibly due to the relatively high solubility of carbon dioxide in water. This might explain why the first experiments with low masses of carbonate and thus small volumes of carbon dioxide give poor results for the estimation of the molar gas volume. This is because a large percentage of the carbon dioxide produced will dissolve in the water. As larger masses of carbonate are used, then a smaller percentage of the carbon dioxide produced will dissolve in the water.

g   From Table 1, estimating the molar gas volume by using a reaction that produces a known amount of hydrogen gas would be better because hydrogen has a much lower solubility in water.

Thus using the same apparatus, but changing the reactants to Mg(s) and H_2SO_4(aq) would be a possible option. Mg ribbon could be coiled in such a way that it sticks at the top of the boiling tube and does not make contact with the acid until the bung is inserted and the reactants mixed by gentle shaking.

An initial experiment should be carried out to ensure that the reaction between the Mg and H_2SO_4 is not too vigorous and amounts should be calculated so that the acid is in excess.

■ Core practical 2

Use a standard solution to find the concentration of a solution of sodium hydroxide

Despite the many recent advances in analytical chemistry, titrimetric analysis still plays an important role in the quantitative analysis of unknown solutions. The basic principle is as follows

■ The volume of a solution of known concentration (the **standard**) is added slowly to a known volume of solution of which the concentration is only approximately known.

■ The end point of the reaction is indicated by a rapid change that can be measured, such as a colour change as determined by an indicator.

In this practical the end point of the titration is determined by the clear colour change of the indicator used. The validity of your results will be determined by the precision to which you make your standard solution. You need to make sure that you are comfortable using the following equation:

$$\text{concentration of solution (mol dm}^{-3}) = \frac{\text{number of moles}}{\text{volume of solution (dm}^3)}$$

Before we consider the titration, it is important that we consider the preparation of the standard solution. The quality of a measurement will depend on the precision of your standard.

Procedure 1: the standard

1 Weigh an empty test tube. Weigh approximately 2.5 g of sulfamic acid into the test tube.

2 Reweigh the test tube and its contents accurately.

3 Add the sulfamic acid to approximately 100 cm³ of distilled water in a 250 cm³ beaker.

4 Reweigh the test tube. The mass difference is the mass of sulfamic acid you have used to make your standard solution.

5 Ensure all of the sulfamic acid has dissolved and then transfer the solution into a 250 cm³ volumetric flask. Ensure that you rinse the beaker and add all washings to the volumetric flask.

6 Make up the final volume to 250 cm³ in the volumetric flask and then stopper the flask.

Introductory questions

What makes a good standard?

A good standard for titrimetric analysis should have the following qualities:

- It should be available at a high level of purity.
- It should be a solid of a high molar mass.
- It should be stable in air (that is, it should not react with carbon dioxide, water vapour or oxygen).
- It should have a high solubility in the solvent (usually water).

Let us consider the standard used in this practical. It is sulfamic acid, NH_2SO_3H. An alternative standard is potassium hydrogen phthalate (KHP for short), $C_8H_5O_4K$ (also an acid).

a Calculate the relative molecular mass of the two standards to 1 d.p.

b Assuming the mass balance used can record to ± 0.01 g, calculate the percentage error for the two standards.

c Which of the two standards offers the greater precision? Explain your answer.

Practical tip

Remember to use the relative atomic masses as given on the periodic table in your data book.

Procedure 2: the titration

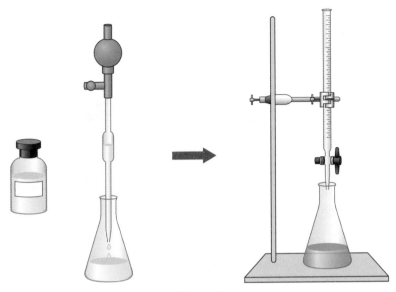

Figure 2

1 Using a suitable volumetric pipette, transfer $25.00\,cm^3$ of the sodium hydroxide into a $250\,cm^3$ conical flask (Figure 2).
2 Add four drops of methyl orange indicator to the conical flask and note the colour.
3 Titrate the contents of the conical flask against your standard. You should aim to carry out a rough titration followed by titrations to a precision of no less than $\pm0.05\,cm^3$.
4 You can stop once you have two concordant titrations.
5 Record all your results in a suitable table.

Sample results

Table 3

Mass of test tube + sulfamic acid/g	21.47
Mass of test tube – sulfamic acid/g	19.05
Mass of sulfamic acid added/g	2.42

Table 4

Volume/cm³	Rough	1st	2nd	3rd
Final burette reading	26.0	23.85	23.95	23.90
Initial burette reading	0.00	0.00	0.00	0.00
Volume added		23.85	23.95	23.90

Analysis of results

a Calculate the concentration of your standard to three significant figures.
b Use your results to calculate a suitable average for your titration.

Practical tip

Remember concordant titres must be within $0.20\,cm^3$ of each other.

c   Write a balanced equation for the reaction.

d   How did you decide when the end point of the reaction was reached?

e   Calculate the concentration of the sodium hydroxide solution to a suitable number of significant figures.

f   The balance used in the experiment is accurate to ±0.01 g. The volumetric flask is accurate to ±0.05 cm³. Calculate the percentage uncertainty in your standard, showing all working.

g   Calculate the percentage error when using the pipette.

h   Calculate the percentage uncertainty in your average titration.

i   Hence, using your answers to parts **g** and **h**, calculate the highest concentration of the sodium hydroxide possible within the range of uncertainty.

j   A student left the top off the sodium hydroxide solution for a period of several hours. Suggest how this might affect the concentration of the sodium hydroxide.

> **Practical tip**
>
> Remember, carbon dioxide is present in the air and as an acidic gas is capable of neutralising an alkali.

Evaluate an alternative procedure

An alternative method for assessing the concentration of sodium hydroxide is proposed by a second student who says that their method is quicker and less prone to error because fewer measurements need to be taken.

1   50 cm³ of sodium hydroxide solution is measured out.

2   An excess of 1.0 mol dm⁻³ magnesium sulfate is added to the sodium hydroxide solution.

3   The precipitate formed is removed by filtration and dried.

4   The mass of the residue is weighed.

The proposed reaction is shown by the following equation:

$$2NaOH(aq) + MgSO_4(aq) \rightarrow Mg(OH)_2(s) + Na_2SO_4(aq)$$

After drying, the mass of the residue collected was 0.13 g.

Alternative procedure questions

a   Use this information to calculate the concentration of the sodium hydroxide solution.

b   Comment on the two procedures to determine the concentration of the sodium hydroxide solution. Which of the two procedures is likely to give the more accurate result? Explain your choice. You should refer to any assumptions you have made in both reactions and possible sources of experimental error, and compare possible percentage errors between the two processes.

c   Calculate the possible range of values from the two experiments.

Answers

Introductory questions

a   Sulfamic acid M_r = 97.1; KHP M_r = 204.2

b   (0.01/97.1) × 100 = 0.01% for sulfamic acid; (0.01/204.2) × 100 = 0.005% for KHP

c   KHP gives greater precision because a larger M_r means a lower percentage error.

Analysis of results

a $2.42/97.1 = 0.0249$ moles. $0.0249/0.250 = 0.0997\,\text{mol dm}^{-3}$ to 3 s.f.

b Average of three accurate titrations $= 23.90\,\text{cm}^3$

c $NH_2SO_3H + NaOH \rightarrow NH_2SO_3^-Na^+ + H_2O$

d The peach colour of the indicator is intermediate between red in alkali and yellow in acid.

e $23.90\,\text{cm}^3$ of 0.0997 standard has 2.38×10^{-3} moles of standard. 1:1 ratio, so concentration of standard $= 2.38 \times 10^{-3}/0.025 = 0.0952\,\text{mol dm}^{-3}$

f Standard uncertainty. Mass $\pm 0.01\%$. Volumetric flask $(\pm 0.05/250) \times 100 = 0.02\%$. Sum of percentage errors $= 0.01 + 0.02 = \pm 0.03\%$

g Percentage error in pipette $= (0.05/25) \times 100 = 0.2\%$

h Titration percentage error $= \pm(0.05 \times 2)/23.9 = \pm 0.42\%$

Sum of all percentage errors $= 0.03 + 0.2 + 0.42 = 0.45\%$

i Maximum concentration of NaOH in range $= 0.0952 + (0.45/100) = 0.0997\,\text{mol m}^{-3}$

j Two possible suggestions: (1) Evaporation of water would increase the concentration of NaOH solution. (2) The reaction of NaOH (aq) with carbon dioxide in the air would reduce the concentration of the NaOH (aq):

$$2NaOH + CO_2 \rightarrow Na_2CO_3 + H_2O$$

Alternative procedure questions

a Assuming that the mass is due to dry $Mg(OH)_2$ $(M_r = 58.3)$, then $0.13\,\text{g} = 0.13/58.3 = 2.23 \times 10^{-3}$ moles. The precipitation reaction can be represented by:

$$Mg^{2+}(aq) + 2NaOH(aq) \rightarrow Mg(OH)_2(s) + 2Na^{2+}(aq)$$

So this is precipitated from $2 \times 2.23 \times 10^{-3}$ moles of NaOH $= 4.46 \times 10^{-3}$ moles, which is in a volume of $50\,\text{cm}^3$, so the concentration of NaOH $= 4.46 \times 10^{-3}/0.05 = 0.089(2)\,\text{mol dm}^{-3}$.

b Points to consider with the second method. The percentage error is greater. Balance $\pm 0.01\,\text{g}$, so the percentage error for mass reading $= (0.01/0.13) \times 100 = 7.7\%$ error. There is a possible mass loss during filtration.

Final value can only be given to 2 s.f., as mass is only given to 2 s.f.

c Experiment 1 range 0.0907–0.0997. Experiment 2 range 0.082–0.096.

■Core practical 3

Find the concentration of a solution of hydrochloric acid

In this second titrimetric analysis we will try to determine a value for the concentration of a solution of approximately $1\,\text{mol dm}^{-3}$ concentration. Note that titrations are not carried out with solutions of concentration as high as $1\,\text{mol dm}^{-3}$ and so an appropriate dilution must be made prior to titrimetric analysis. In this titration a different indicator is used.

Procedure

1 Transfer 25.0 cm³ of hydrochloric acid solution (approximate concentration 1 mol dm⁻³) using a volumetric pipette into a clean 250 cm³ volumetric flask and make up the total volume to 250 cm³ using distilled water.

2 Transfer 25.00 cm³ of this solution into a conical flask and add two to three drops of phenolphthalein indicator.

3 Titrate the contents of the flask against the standardised sodium hydroxide solution until the end point is reached.

4 Repeat until concordant titrations are achieved.

Sample results

Table 5

Final burette reading/cm³	24	24.35	26.45	26.35	26.35
Initial burette reading/cm³	0	0.00	2.10	0.00	2.00
Titre/cm³	24				

Sodium hydroxide standardised at 0.096 mol dm⁻³

Results analysis

a Complete the results table and calculate a suitable average titre.

b Write a balanced full and ionic equation for the reaction.

c Calculate the *original* concentration of the hydrochloric acid to the appropriate number of significant figures.

d Suggest why the end point shown in Figure 3 is unsuitable as an accurate end point.

e A set of class results consistently indicate a concentration value for the hydrochloric acid lower than the true value. Is this likely to be due to systematic or random error? Explain your answer and suggest a reason for the error in this case.

f Five minutes after the end point has been reached, the pink colour in the conical flask has disappeared. With the aid of equations, suggest why this might be the case.

Research suggests that the compound hydrated sodium tetraborate would be a better standard with which to standardise a solution of hydrochloric acid.

g Balance the equation for the reaction of hydrated sodium tetraborate with hydrochloric acid using the appropriate stoichiometric coefficients:

$$Na_2B_4O_7.10H_2O + HCl \rightarrow H_3BO_3 + NaCl + H_2O$$

h Show by calculation what mass of $Na_2B_4O_7.10H_2O$ is required to make 500 cm³ of a 0.0125 molar solution.

Practical tip

Remember to allow for the dilution by a factor of 10 at the start of the procedure. Many students get caught out by this in exam questions.

Figure 3

Answers

Results analysis

a

Final burette reading/cm³	24	24.35	26.45	26.35	26.35
Initial burette reading/cm³	0	0.00	2.10	0.00	2.00
Titre/cm³	24	**24.35**	**24.35**	26.35	**24.35**

Average should be based on titres 2,3 and 5. This gives $24.35\,cm^3$.

b Full: $NaOH(aq) + HCl(aq) \rightarrow NaCl(aq) + H_2O(l)$

Ionic: $OH^-(aq) + H^+(aq) \rightarrow H_2O(l)$

c $0.02435 \times 0.096 = 2.338 \times 10^{-3}/0.025 = 0.0935$, so original concentration $\times 10 = 0.94\,mol\,dm^{-3}$ to 2 s.f. (standard is only given to 2 s.f.)

d The colour change is too strong and thus reaction is well past the end point.

e It is likely to be due to systematic error, as repeating the experiment will generate random errors: that is, favouring values either above or below the true value. It could be due to the preparation of an inaccurate standard used by the whole class.

f Carbon dioxide in the air begins to neutralise the very small excess of NaOH at the end point.

$2NaOH + CO_2 \rightarrow Na_2CO_3 + H_2O$

g $Na_2B_4O_7.10H_2O + \mathbf{2}HCl \rightarrow \mathbf{4}H_3BO_3 + \mathbf{2}NaCl + \mathbf{5}H_2O$

h $M_r = 381.4$; $0.5 \times 0.0125 = 0.00625$ moles needed; $2.38\,g$

■ Core practical 4

Investigation of the rates of the hydrolysis of halogenoalkanes

Introduction

Neutralisation reactions (see CP2 and CP3) involve interactions between ionic species and thus tend to happen very quickly. Reactions involving covalent bonds generally have higher activation energies and thus take longer. In the first part of this experiment we will compare the rate of hydrolysis of chloro, bromo and iodobutane. The solvent used is aqueous ethanol as halogenoalkanes are not sufficiently soluble in pure water, but the mixture of the ethanolic solution and the aqueous silver nitrate is sufficiently miscible to allow the reaction to take place. In the second part of the experiment we will compare the rates of reaction between water and three isomers of bromobutane: 1-bromobutane, 2-bromobutane and 2-bromo-2-methylpropane.

At GCSE you will have learned that for a reaction to take place between two reactant molecules, atoms or ions, two conditions must be met: (i) the particles must collide and (ii) the particles must collide with sufficient energy. At A-level you will see that many reactions are more complicated than this and may take place in a number of steps in sequence. Investigations into **reaction rates** will help us understand what is happening at the molecular level of a reaction and enable us to propose **reaction mechanisms**. You will investigate this further in CP13.

Practical tip

Make sure that you know the general form of a rate equation.

Introductory questions

a Draw a dot-and-cross diagram of a water molecule showing:

    i all valence electrons

    ii the HOH bond angle

    iii any molecular dipole

b Explain why a water molecule can act as a **nucleophile**.

c The reactions only take place at a reasonable rate if they take place in a water bath at 50°C. What does this suggest about the activation energy for this type of reaction in comparison with a precipitation reaction?

d Draw skeletal structures for the three isomers of bromobutane named above. Suggest a modern instrumental analytical technique that would enable you to distinguish between the three isomers and give details of the results that would enable you to make the distinctions. (You might be asked to sketch such a spectrum in an exam question.)

e A fourth isomer of bromobutane is a **primary** bromoalkane and its proton NMR spectrum shows three distinct proton environments. Use this information to draw this isomer, name it, and account for the different proton environments.

> A **nucleophile** is an ion or a molecule that can donate an electron pair.

> **Practical tip**
>
> You should be familiar with instrumental analytical techniques, including NMR, IR and mass spectrometry.

Procedure: part 1

1 Fill a $250\,cm^3$ beaker to about three-quarters full with water at 50°C.

2 Fill three test tubes with $5\,cm^3$ of ethanol.

3 Add four drops of 1-chlorobutane to the first of the three test tubes, four drops of 1-bromobutane to the second and four drops of 1-iodobutane to the third, and place a bung loosely on each test tube.

4 Pour $5\,cm^3$ of aqueous silver nitrate into three more test tubes and place them in the water bath.

5 Allow all six test tubes to warm in the beaker for about 5 minutes

6 Transfer one of the $5\,cm^3$ volumes of aqueous silver nitrate into the 1-chlorobutane test tube and immediately start the stopwatch.

7 Record the time taken for the mixture to become cloudy to the nearest second.

8 Repeat steps 6 and 7 for the remaining two halogenoalkanes.

Sample results

Table 6

Halogenoalkane	Time taken for precipitate to form/s	C–X bond enthalpy/kJ mol^{-1}
1-chlorobutane	595	
1-bromobutane	82	
1-iodobutane	48	

Procedure: part 2

Repeat the procedure outlined in part 1 for the compounds 1-bromobutane, 2-bromobutane and 2-bromo-2-methylpropane.

Sample results

Table 7

Halogenoalkane	Time taken for precipitate to form/s
1-bromobutane	78
2-bromobutane	38
2-bromo-2-methylpropane	6

Questions on part 1

a Using your reference data booklet, look up bond enthalpy values for the C–X bonds and complete the column in the sample results table. Use these data to explain the experimental results shown in Table 6.

b Draw a full reaction mechanism for the reaction between a water molecule and 1-bromobutane. Be sure to show all relevant electron pairs, curly arrows and bond dipoles.

c Write an equation, including state symbols, for the formation of the precipitate in part **b**.

d Explain what is meant by heterolytic fission and suggest why the precipitate forms *as soon as* the C–Br bond breaks.

e Suggest *two* improvements that could be made to improve the reliability of this investigation.

Questions on part 2

f Investigations into the rate of the reaction between water and 2-bromo-2-methylpropane show that the full **rate equation** has the following form:

rate = $k[(CH_3)_3CBr]$

i Explain why this is consistent with a two-step reaction mechanism and suggest which step is the **rate-determining step** for the reaction.

ii Sketch an energy profile diagram that is consistent with the rate equation shown above. On your sketch mark any activation energies and intermediate structures. You can assume that the overall reaction is exothermic.

Answers

Introductory questions

a

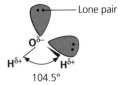

Practical tip

X in this context signifies any of the possible halogens in the experiment.

b A water molecule has two lone pairs of electrons, either of which can be donated to form a covalent bond.

c It has a reasonably high activation energy in comparison with a precipitation reaction.

d

Possible answers:

(i) Proton NMR. There are different numbers of proton environments: One bromo would have four environments with integral ratios 3,2,2,2. Two bromos would have four environments with integral ratios 3,2,1,3. 2-bromo-2-methylpropane would have only one proton environment (nine equivalent protons).

(ii) Mass spectrometry could also be used. Only one bromo would give a $[CH_2Br]^+$ peak. Only two bromos would give $[CH_3CH_2CBr]^+$. Only 2-bromo-2-methylpropane would give a $[(CH_3)_3C]^+$ peak.

There are other possible examples. See: http://sdbs.db.aist.go.jp/sdbs/cgi-bin/cre_index.cgi

e The fourth isomer is 1-bromo-2-methylpropane. Three different proton environments are shown using **a**, **b** and **c** in Figure 4.

Note that integrals would be **a** = 6, **b** = 1 and **c** = 2.

Figure 4

Answers to questions on parts 1 and 2

a

C–X	Energy/kJ mol⁻¹
C–Cl	346
C–Br	290
C–I	228

Higher reaction rate for lower C–X bond enthalpy

b

c $Ag^+(aq) + Br^-(aq) \rightarrow AgBr(s)$

d   Heterolytic fission is the breaking of a covalent bond such that the two fragments are oppositely charged ions. Once the bond breaks heterolytically, the Br^- is released into solution where it immediately attracts the Ag^+ ion and forms a precipitate of AgBr.

e   Suggested improvements: (i) Four drops is an equal volume of the halogenoalkane, but equimolar quantities would be a better measure of quantity. (ii) There is difficulty in measuring when a precipitate appears. Suggest a light meter to measure quantitative absorbance when a specific amount of precipitate has been formed.

f   i   The rate equation has only $(CH_3)_3CBr$ in it, suggesting that only this molecule is in the rate determining step. Therefore there must be a second fast step if the product is to form.

   ii

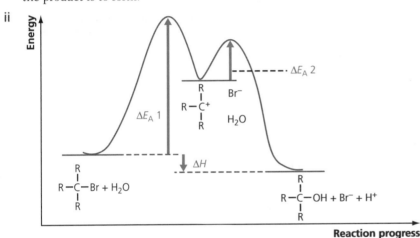

■ Core practical 5

Investigation of the oxidation of ethanol

Introduction

Organic chemistry is essentially the chemistry of carbon and its compounds, but such is the versatility of carbon that there is effectively an infinite number of carbon-based molecules. There are, for example, nearly 37 million different molecular structures with the formula $C_{25}H_{52}$.

A reaction as seemingly straightforward as the oxidation of ethanol is more complicated than you might expect. Consider that the complete oxidation of ethanol (or combustion) leads to two products — carbon dioxide and water — but a more controlled oxidation can yield ethanal, ethanoic acid or a mixture of the two.

In this practical we will be looking at the oxidation of ethanol to ethanoic acid and how reagents and conditions are controlled in such a way as to maximise the yield of ethanoic acid. The **oxidising agent** in this reaction is acidified sodium dichromate. The half-equation for the reaction is shown in equation 1.

$$14H^+(aq) + Cr_2O_7{}^{2-}(aq) + 6e^- \rightarrow 2Cr^{3+}(aq) + 7H_2O(l) \qquad \textit{Equation 1}$$

Once you have carried out the organic synthesis, you will carry out a series of qualitative tests, record observations and then make deductions about your distillate.

Introductory questions

a Assign oxidation numbers to the chromium species in equation 1.

b Why is equation 1 referred to as a 'half-reaction'?

$$C_2H_5OH(l) + H_2O(l) \rightarrow CH_3COOH(aq) + 4H^+(aq) + 4e^- \qquad \textit{Equation 2}$$

c How can you tell that equation 2 represents an oxidation process?

d Combine the two half-equations, 1 and 2, to give the overall ionic equation for the reaction.

e Calculate the *minimum* volume of $1.70 \, mol \, dm^{-3}$ acidified sodium dichromate required to convert 2.0 g of ethanol to ethanoic acid.

Procedure

1 Add $10 \, cm^3$ of acidified sodium dichromate solution to a $50 \, cm^3$ pear-shaped flask.

2 Add a few anti-bumping granules to the pear-shaped flask.

3 Set up the flask for heating under reflux as shown in Figure 5.

4 Mix $2 \, cm^3$ of ethanol with $5 \, cm^3$ of distilled water in a boiling tube.

5 Add the ethanol solution dropwise down the reflux condenser into the pear-shaped flask.

6 Once all of the ethanol solution has been added, heat the contents of the pear-shaped flask carefully for about 30 minutes using a micro-burner.

7 Once the apparatus has cooled sufficiently that it can be comfortably held by hand, re-configure your apparatus for distillation and collect the products that boil between 114°C and 120°C (see Figure 6).

Figure 5

Figure 6

An **oxidising agent** is a substance that brings about oxidation in other substances by being itself reduced. Oxidising agents are compounds in which elements are assigned high oxidation numbers. For example, Mn has a +7 oxidation number in $KMnO_4$.

Practical tip

When you are asked to give the name of an oxidising agent 'acidified dichromate' is not sufficient. You must give the full name of the compound.

Tests on distillate

Split the distillate into four approximately equal volumes of about $1\,cm^3$ and then carry out the following tests:

a Measure the pH of the distillate using universal indicator paper.

b Add a few drops of acidified potassium dichromate to the $1\,cm^3$ volume of the distillate used in part **a** and warm the solution in a water bath at 60°C.

c Add a quarter of a spatula of calcium carbonate to a $1\,cm^3$ volume of the distillate.

d Add a 1 cm length piece of clean magnesium ribbon to the third $1\,cm^3$ of the distillate.

e Add $1\,cm^3$ of Fehling's solution to the final $1\,cm^3$ of the distillate and gently warm using a Bunsen burner.

Sample results

Table 8

Test carried out	Observations
pH of distillate	Orange colour equal to pH 3
Acidified potassium dichromate added	No observable change
Calcium carbonate added	Slight fizzing, cloudy solution
Magnesium ribbon added	Slight fizzing
Fehling's solution added	Clear blue solution to slight cloudy green colour

Questions on results

a Using an appropriate structural formula and equations, explain why the distillate has a pH of 3.

b Write out an **ionic** equation for the reaction between calcium carbonate and the distillate and include state symbols.

c Using two appropriate **half-equations** explain why the reaction between magnesium ribbon and the distillate can be classified as a **redox** reaction.

d Comment on the result of the reaction between the distillate and the Fehling's solution; what is suggested? (Clue: look back at your answer to question **e** in the introductory questions and compare it with the volume used in the core practical instructions.)

e i Draw full displayed formula for the molecules ethanol, ethanal, and ethanoic acid.

   ii Identify which of the three molecules that you have drawn in part **i** cannot form hydrogen bonds with itself and explain why this is the case.

   iii Given your answer to part **ii**, how might you modify the procedure outlined in this core practical to synthesise ethanal from ethanol rather than ethanoic acid?

Practical tip

It is important to remember *not to expect a positive result* when you carry out a test. A true negative result is every bit as important when making your inferences as a true positive.

Practical tip

This type of question may form part of an extended question in an exam. You should plan your answer in rough first and transfer to the script only when you are happy that you have sufficient detail. Be sure to name all apparatus and reagents in full.

Answers

Introductory questions

a +6 in $Cr_2O_7^{2-}$(aq) and +3 in Cr^{3+}(aq)

b The reaction shows only the oxidation process and as both oxidation and reduction have to occur simultaneously, only half of the reaction is shown.

c Electrons are 'lost' (on the product side of the half-equation).

d $3C_2H_5OH + 2Cr_2O_7^{2-} + 16H^+(aq) \rightarrow 3CH_3COOH(aq) + 4Cr^{3+}(aq) + 11H_2O(l)$

e 2.00 g of ethanol = 2/46 = 0.0435 moles of ethanol. Moles of dichromate required from equation = (2/3) × 0.0435 = 0.0290 moles of dichromate. If concentration = 1.70, then volume = 0.029/1.7 = 0.0171 dm^3 or 17.1 cm^3. (We will come back to this answer later in answers to questions on results **d**.)

Questions on results

a

Ethanoic acid is a weak acid and donates an H^+ ion to a water molecule, forming the H_3O^+ ion, which causes acidity.

b $CaCO_3(s) + 2H^+(aq) \rightarrow Ca^{2+}(aq) + H_2O(l) + CO_2(g)$

c $Mg(s) \rightarrow Mg^{2+}(aq) + 2e^-$ (oxidation)

$2H^+(aq) + 2e^- \rightarrow H_2(g)$ (reduction)

Overall reaction: magnesium is oxidised; hydrogen ions are reduced.

d Slight reaction (cloudy green), suggesting there is something in the mixture to be oxidised. Now look back at your answer to question **e** in the introductory questions and the reaction procedure.

e i

Ethanol   Ethanal   Ethanoic acid

ii Ethanal cannot form hydrogen bonds with itself. It does not have an H atom directly covalently bonded to N, O or F.

iii If ethanal does not form hydrogen bonds with itself, then it is likely to have a lower boiling point. If apparatus is set up for distillation and oxidising agent is added *dropwise*, then as ethanal is formed (boiling point 21°C) it will immediately boil and be condensed. No reflux condenser means that the ethanal will not have the chance to be condensed, returned to the reaction flask and oxidised further.

Core practical 6

Chlorination of 2-methylpropan-2-ol using concentrated hydrochloric acid

Introduction

We have already seen that carbon-based compounds can form a vast number of different structures. When we are carrying out a *synthesis* reaction this can make life difficult as there may be a number of side reactions leading to more than the one product we want. As a result, synthesis reactions have to be carried out carefully and the purification process at the end of the reaction is extremely important. Exactly how pure a product must be will depend upon its use: for example, the purity of a medicine must be as close to 100% as possible; for a paint solvent purity is less important.

In this reaction you will carry out the conversion of 2-methylpropan-2-ol into 2-chloro-2-methylpropane. On completion of the synthesis you will purify the sample. You will also be asked to consider how modern instrumental techniques can be used to analyse the quality of your sample.

Procedure

1 Pour $20\,cm^3$ of 2-methylpropan-2-ol and $70\,cm^3$ of concentrated hydrochloric acid into a large conical flask. Place a bung on the flask and mix the contents by swirling.

2 After about 30 seconds stop swirling and carefully ease off the bung to release any pressure build-up.

3 Continue steps 1 and 2 for about 20 minutes, being careful to release the pressure regularly.

4 After about 20 minutes the contents of the flask will have divided into two layers or phases.

5 Now add about 6 g of anhydrous calcium chloride to the mixture and swirl until it has dissolved. The calcium chloride will dissolve in the aqueous layer but not in the chloroalkane layer

6 Transfer the contents of the conical flask into a separating funnel and allow the two layers to separate. You can now carefully run off and discard the lower aqueous layer and retain the organic layer.

7 Add $20\,cm^3$ of sodium hydrogen carbonate to the separating funnel and place a bung in the top.

8 Mix the contents by inverting the funnel, taking care to release the pressure with each mix.

9 Remove the bung from the separating funnel and discard the lower aqueous layer.

10 Repeat steps 7–9 until no more gas is released.

11 Pour the organic layer into a small conical flask and add a spatula-full of anhydrous sodium sulfate and swirl. Leave the sample overnight.

12 Once the liquid is clear, decant the liquid into a $50\,cm^3$ round-bottomed flask and distil the crude product (Figure 7), collecting the fraction that boils between 50 and 52°C.

13 Place the distillate in a sealed vial and label it.

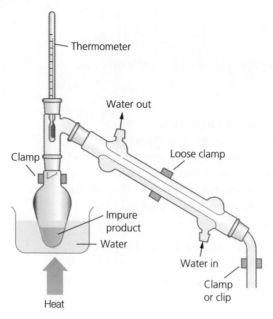

Figure 7

Questions

a The density of 2-methylpropan-2-ol is $0.775\,g\,cm^{-3}$. Calculate the number of moles in $20\,cm^3$ of this liquid.

b i $70\,cm^3$ of concentrated hydrochloric acid ($11.65\,mol\,dm^{-3}$) is used in step 1 of this reaction. Calculate the number of moles of HCl.

   ii Suggest why the acid is used in excess in this reaction.

c In step 7, sodium hydrogen carbonate solution is added to neutralise any excess acid. Write an **ionic** equation for this reaction and identify the gas produced.

d Figure 8 shows an IR spectrum of the product formed in this reaction.

Practical tip

This reaction is unlikely to go to completion but rather will find a position of equilibrium.

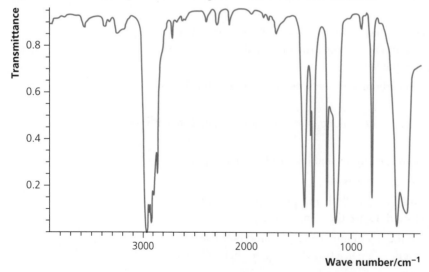

Figure 8 Infrared spectrum for 2-chloro-2-methylpropane

i How can this spectrum be used to assess the purity of the product?

ii What might a small but broad absorbance centred around 3400 cm^{-1} suggest about your product?

e What would be the most distinctive feature of the ^{1}H NMR spectrum of the pure product in this case?

f The low-resolution mass spectrum of the product gives *two* molecular ion peaks at *m/z* values of 92 and 94. These peaks are in a ratio of approximately 3:1 in height. Explain this observation.

g An investigation into the kinetics of this nucleophilic substitution reaction establishes the following rate equation:

rate = $k[(CH_3)_3COH]$

Using this information, draw a reaction mechanism that is consistent with this rate equation.

Answers

a 20 cm^3 has a mass of 20 × 0.775 = 15.5 g. Number of moles = 15.5/M_r of 2-methylpropan-2-ol (74 g mol^{-1}), so 15.5/74 = 0.210 moles.

b i Concentration × volume in dm^3 = number of moles. Thus 11.65 × 0.07 = 0.816 moles.

ii A higher concentration of acid is likely to shift the equilibrium to the right-hand side and ensure that most of the 2-methylpropan-2-ol is converted.

c $H^+(aq) + HCO_3^-(aq) \rightarrow H_2O(l) + CO_2(g)$

Carbon dioxide gas is produced.

d i Compare the IR spectrum with a database for a match. Peaks in 'fingerprint' region should be a good match with a reliable database.

ii This would suggest an OH group absorbance that could be due to either unconverted alcohol or water in the sample.

e The ^{1}H NMR spectrum would be distinctive in that a pure product would only have one proton environment as 2-chloro-2-methylpropane has nine equivalent protons. The peak would be somewhere between 0.0 and 2.0 ppm.

f The 92 molecular ion peak is due to an M+ ion with a ^{35}Cl isotope. The 94 is due to an M+ ion with a ^{37}Cl atom. The ratio of 3:1 reflects the 3:1 abundance ratio of the two chlorine isotopes.

g

The reaction mechanism is SN_1, as only the 2-methylpropan-2-ol is in the rate equation. This rate-determining step must involve only 2-methylpropan-2-ol.

■ Core practical 7

Analysis of some inorganic and organic unknowns

Introduction

The job of the analytical chemist has never been more important than it is today. In areas as diverse as pharmaceuticals, food science, forensics and environmental science, to name but four, the analytical chemist plays a key role. There are essentially two types of analysis: **qualitative** and **quantitative**. Put simply, qualitative = what is there? And quantitative = how much is there? In some cases it is simply important to know that the amount of a substance does not rise above a given threshold; for example, levels of nitrate below $50\,mg\,dm^{-3}$ in drinking water are considered safe by the World Health Organization (WHO) (most tap water in the UK has a significantly lower value than this).

Much analytical chemistry is now carried out using complex instrumental techniques (many of which you will study in this course), but there is still a role for the 'wet' laboratory chemical tests. In this investigation you will be asked to identify *both the anion and the cation* in the inorganic samples labelled X, Y and Z. In this analysis you can assume that there is *only one anion and one cation* in each sample. For the organic samples labelled A, B and C, you will be asked to identify the one **functional group** present in each sample.

> **Practical tip**
>
> The inorganic compounds are solids at room temperature, whereas the organic samples are liquids. You should be able to explain why this is in terms of structure and bonding.

Procedure: part 1

In investigations such as this, it is important to have a results table prepared. Read through the instructions carefully first and then prepare an appropriate table. Be careful to observe any reactions you see and leave the inferences until after the experiment.

Perform each of the following tests on samples A, B and C. Each of the organic liquids has four carbon atoms per molecule and one functional group.

1 Place ten drops of each liquid into three separate test tubes, add a 1 cm depth of bromine water to each.

2 Place ten drops of each liquid into three separate test tubes. Add $1\,cm^3$ of acidified potassium dichromate solution to each and then warm in a 60°C water bath for about 5 minutes.

3 Place ten drops of each liquid into three separate test tubes. Add $1\,cm^3$ of Fehling's solution to each test tube and heat in a water bath.

4 Place ten drops of each liquid into three separate test tubes. Add $1\,cm^3$ of ethanol and $1\,cm^3$ of sodium hydroxide solution to each test tube and warm in a water bath for about 5 minutes. After 5 minutes remove the test tubes from the water bath, add a few drops of nitric acid and then add five drops of silver nitrate solution.

> **Practical tip**
>
> Frequently information given at the beginning of the question is important to later parts of the question. Don't forget to refer back to the introductory stem of the question.

Procedure: part 2

Perform the following tests on each of the inorganic solids X, Y and Z.

1 Carry out flame tests on a sample of each of the three solids.

2 Dissolve a spatula-full of each solid in three separate test tubes containing $10\,cm^3$ of distilled water. Split each solution into three roughly equal portions and then carry out the following tests:

   i To the first portion of each add $5\,cm^3$ of dilute nitric acid followed by ten drops of silver nitrate solution. Then add dilute ammonia solution dropwise.

   ii To the second portions, add $5\,cm^3$ of dilute nitric acid followed by ten drops of barium chloride solution.

   iii To the third portions add $2\,cm^3$ of chlorine water.

Sample results: part 1

Table 9

Test/sample and observations	A	B	C
Bromine test	Solution remained orange. Two layers formed.	Solution turned from orange to colourless. Two layers formed.	Solution remained orange.
Acidified potassium dichromate test	Solution remained orange.	Solution remained orange.	After 3 minutes solution became a brown/green colour.
Fehling's solution test	Blue Fehling's solution did not change.	Blue Fehling's solution did not change.	Blue Fehling's solution did not change.
Ethanol, sodium hydroxide, nitric acid and silver nitrate test	After silver nitrate was added, a pale cream precipitate formed. Precipitate still present after ammonia solution added.	A small amount of greyish precipitate at bottom of test tube. Dissolved when ammonia solution was added.	No changes observed.

Questions: part 1

a When bromine water is added to liquid A, two separate layers form. What does this suggest about liquid A?

b Draw the mechanism for the reaction that explains the observation between bromine water and liquid B. You can use R groups for alkyl groups where necessary. Give a name for type of reaction taking place.

c What does the result in the second test for liquid C suggest?

d What does Fehling's solution test for and what can we conclude from the three negative results?

e In carrying out the final test on liquid B a student thought that they might have forgotten to add the nitric acid. Can you suggest an identity for the grey precipitate and why it formed?

f Write an ionic equation for the formation of the precipitate in the last test on liquid A.

g Suggest possible identities for liquids A, B and C that are consistent with the information above and your observations.

Practical tip

You must ensure that when drawing mechanisms the arrows are drawn pointing the right way: remember that an arrow shows the movement of an electron pair.

Core Practicals

Sample results: part 2

Table 10

Test/sample and observations	X	Y	Z
Flame test colour	Crimson red	Orange yellow	Lilac
Nitric acid, silver nitrate and ammonia solution	Some bubbles when nitric acid added. No further reaction.	Pale yellow precipitate with silver nitrate. No change with ammonia solution.	No reactions observed.
Nitric acid and barium chloride solution	Some bubbles when nitric acid added. No further reactions.	No reactions observed.	White precipitate formed on addition of barium chloride.
Chlorine water	No reaction observed.	Solution turned a red/brown colour.	No reactions observed.

Questions: part 2

a Identify the cations in samples X, Y and Z.

b Write an ionic equation, including state symbols for the reaction between a solution of Y and silver nitrate solution.

c Write an ionic equation for the formation of the white precipitate in the third test with a solution of Z.

d Write an ionic equation for the reaction that gives a red/brown solution when the fourth test is carried out on Y and identify the species responsible for the colour.

e Give the formulae for X, Y and Z.

Answers

Questions: part 1

a A is immiscible with water. So it is unlikely to be an alcohol. (It cannot form hydrogen bonds.)

b

c This suggests that C is either a primary or secondary alcohol or possibly an aldehyde, as the acidified potassium dichromate is being reduced.

d Fehling's solution tests for an aldehyde functional group. We can thus conclude that none of the liquids is an aldehyde, and thus C must be a primary or secondary alcohol.

e The precipitate is probably silver hydroxide AgOH(s). If nitric acid is not added, then the sodium hydroxide is not neutralised.

f $Ag^+(aq) + Br^-(aq) \rightarrow AgBr(s)$

g   A must have four carbons and a bromine. It is any isomer of bromobutane. B must have four carbons and at least one C=C bond. It is an isomer of butene, or possibly a butadiene. C must have four carbons and have a primary or secondary alcohol group. It is butan-1-ol or butan-2-ol or 2-methylpropan-1-ol.

Part 2

a   Cations present in X = Li; Y = Na; Z = K

b $Ag^+(aq) + I^-(aq) \rightarrow AgI(s)$

c $Ba^{2+}(aq) + SO_4^{2-}(aq) \rightarrow BaSO_4(s)$

d $Cl_2(aq) + 2I^-(aq) \rightarrow I_2(aq) + 2Cl^-(aq)$

Iodine in solution causes the red/brown colour.

e $X = Li_2CO_3; Y = NaI; Z = K_2SO_4$

■ Core practical 8

To determine the enthalpy change of reaction using Hess's law

Introduction

One of the most fundamental ideas in the physical sciences is summed up in the first law of thermodynamics, which, simply put, states that the amount of energy in the universe remains constant. When we talk of 'running out of energy' what we mean is that we are running out of useful energy, the kind that can be used to do work. **Hess's law** is an example of the first law of thermodynamics applied to chemical systems.

Hess's law states: 'The energy change for a reaction is independent of the route taken for the reaction.'

Thus for the reaction scheme in Figure 9 we can say that $\Delta H_1 = \Delta H_2 + \Delta H_3$.

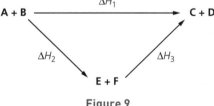

Figure 9

Measuring the energy changes of some reactions directly can be very difficult, especially when you have to continuously supply energy for the reaction to keep going, as with most endothermic reactions. Applying Hess's law means that we can select an alternative route for the same *overall* reaction that is easier to do, knowing that the enthalpy change for the reaction will be the same.

In this practical we are going to measure the enthalpy change for the decomposition of potassium hydrogen carbonate, which is represented in equation 1.

$$2KHCO_3(s) \rightarrow K_2CO_3(s) + CO_2(g) + H_2O(l) \qquad \textit{Equation 1}$$

Practical tip

Always remember to take note of the number of moles in the equation. In this case because 2 moles of potassium hydrogen carbonate are decomposed, you must include a ×2 factor when using the enthalpy of reaction for potassium hydrogen carbonate with acid.

Procedure: part 1

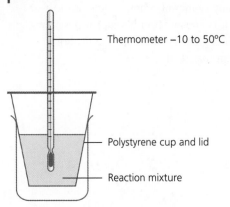

Thermometer −10 to 50°C

Polystyrene cup and lid

Reaction mixture

Figure 10

1 Place approximately 3 g of potassium carbonate into a test tube. Accurately weigh the test tube and its contents to 2 decimal places.

2 Use a burette to add 30 cm³ of 2 mol dm⁻³ hydrochloric acid into a polystyrene cup and support the cup in a beaker (Figure 10).

3 Measure and record the temperature.

4 Now add the potassium carbonate gradually while stirring with the thermometer.

5 Record the maximum temperature achieved once all of the potassium carbonate has been added.

6 Reweigh the empty test tube.

7 Repeat steps 1–6 using approximately 3.5 g of potassium hydrogen carbonate in place of potassium carbonate. Record the *lowest* temperature reached.

Sample data

Table 11

Experiment 1	Mass/g
Test tube + potassium carbonate	11.46
Test tube after potassium carbonate added	8.41
Mass added	3.05
Final temperature	27.5
Initial temperature	20.0
ΔT	(+)7.5

Table 12

Experiment 2	Mass/g
Test tube + potassium hydrogen carbonate	12.97
Test tube after potassium hydrogen carbonate added	9.44
Mass added	3.53
Final temperature	12.0
Initial temperature	20.0
ΔT	(−)8.0

Questions

For the following questions you may make the following assumptions: $1\,cm^3$ of any aqueous solution has a mass of $1\,g$. The **specific heat capacity** of all aqueous solutions $= 4.18\,J\,g^{-1}\,K^{-1}$. Remember that ΔT is the same in °C as in K.

a Write a balanced equation, including state symbols, for the reaction between hydrochloric acid and potassium carbonate.

b Calculate the amount of heat energy transferred to the solution in this experiment.

c Calculate a value for $\Delta_r H_1$ for this reaction, giving your answer to the appropriate number of significant figures and making the units clear.

d Repeat steps **a–c** for the reaction of potassium hydrogen carbonate with hydrochloric acid.

e Construct a Hess's law cycle for the reaction shown in equation 3 from the reactions you have carried out in experiments 1 and 2.

f Hence calculate a $\Delta_r H_2$ value for the thermal decomposition of potassium hydrogen carbonate to potassium carbonate.

g One of the assumptions that we made was that the specific heat capacity of the solutions is the same as that of pure water. Calculate the number of moles of HCl in $30\,cm^3$ of $2.0\,mol\,dm^{-3}$ of water.

h Assuming that there is no increase in volume when HCl gas is dissolved in water, calculate the number of moles of water in $30\,cm^3$ of $2.0\,mol\,dm^{-3}$.

i On the basis of your answers to parts **f** and **g**, do you think your assumption is reasonable?

The **specific heat capacity** of a material is the energy transferred in joules per unit mass (usually per gram) when the material rises in temperature by one degree kelvin. So, for example, it takes $4.18\,J$ to raise $1\,gram$ of water by one degree kelvin.

Answers

a $K_2CO_3(s) + 2HCl(aq) \rightarrow 2KCl(aq) + CO_2(g) + H_2O(l)$

b $Q = 30 \times 4.18 \times 7.5 = 940.5\,J$

c M_r of $K_2CO_3 = 138.2$. Number of moles used $= 3.05/138.2 = 0.02207$. Hence $\Delta_r H = 940.5/0.02207 = 42\,614\,J\,mol^{-1}$ so $\Delta_r H = -43\,kJ\,mol^{-1}$ (to 2 s.f.)

d $KHCO_3(s) + HCl(aq) \rightarrow KCl(aq) + CO_2(g) + H_2O(l)$

$Q = 30 \times 4.18 \times 8.0 = 1003.2\,J$. M_r $KHCO_3 = 100.1$. Moles $= 3.53/100.1 = 0.0353$

So $1003.2/0.0353 = 28\,419\,J\,mol^{-1}$ so $\Delta_r H = +28\,kJ\,mol^{-1}$ (to 2 s.f.)

e

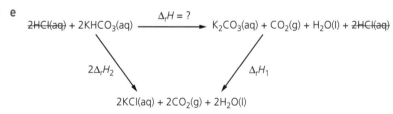

f Hence $\Delta_r H = 2 \times (28) + (+43) = +99\,kJ$. (If you are going to say $kJ\,mol^{-1}$ here, you must make it clear that it is *per mole of potassium carbonate formed*.)

g $30\,cm^3$ of $2.0\,mol\,dm^{-3}$ $HCl(aq)$ is $0.06\,moles$ of HCl.

h $30\,cm^3$ of solution is approximately $30\,g$ of water, which is $30/18 = 1.67$ moles of water.

i Ratio of $HCl:H_2O$ is $0.06:1.67$ or nearly $1:28$. Thus water is the majority component of the solution and thus the approximation is reasonable.

■Core practical 9

Finding the K_a value for a weak acid

Introduction

Acid–base chemistry is an important part of any A-level chemistry course, but is frequently presented in a rather abstract way. It should be noted that many chemical reactions taking place in an aqueous environment can be significantly influenced by pH values (and thus many organic synthesis reactions must take place in buffered solutions). When it comes to biological systems, the control of pH within certain limits is vital and the involvement of weak acids in buffer systems is central to this role; consider the role of carbonic acid in the blood. There will be some questions that test your understanding of buffer systems at the end of the questions on the sample data.

In this practical you will investigate a procedure for experimentally determining the K_a value for a given weak acid. You will be using a pH meter to record values as measured quantities of alkali are added to the acid.

Procedure

1 Ensure that your pH meter is first calibrated using a pH 3.0 buffer solution.

2 Pipette $25.0\,cm^3$ of $0.1\,mol\,dm^{-3}$ into a conical flask.

3 Record the pH of the $0.1\,mol\,dm^{-3}$ to 1 or 2 d.p. where possible.

4 Fill a burette with $0.1\,mol\,dm^{-3}$ sodium hydroxide solution.

5 Add $5.0\,cm^3$ of the sodium hydroxide to the conical flask, stir the contents with the pH meter and record the pH value.

6 Add a further $5.0\,cm^3$ of the sodium hydroxide, stir the contents with the pH meter and record the pH value.

7 Repeat step 6 until a total volume of $20\,cm^3$ of sodium hydroxide has been added.

8 Now repeat step 6 but this time make $1\,cm^3$ additions of sodium hydroxide and record the pH value.

9 Once the recorded pH value goes above 7, take the pH meter out and recalibrate the pH meter using a pH = 10.0 buffer solution.

10 Continue making $1\,cm^3$ additions of sodium hydroxide until a total of $30\,cm^3$ of alkali has been added, making sure the pH value is recorded with each addition.

11 Now make $5\,cm^3$ additions of sodium hydroxide until a total of $50\,cm^3$ of alkali has been added, giving a total volume of $75\,cm^3$.

Sample data

Table 13

Volume of 0.1 mol dm^{-3} NaOH(aq) added/cm^3	pH
0	2.95
5	3.80
10	4.40
15	4.60
20	4.80
21	4.90
22	5.00
23	5.20
24	6.20
25	10.60*
26	11.60
27	11.80
28	12.10
29	12.20
30	12.30
35	12.50
40	12.70
45	12.80
50	12.80

* pH meter recalibrated in pH = 10.0 buffer solution

Questions on data

a   Plot a graph of pH (y-axis) versus volume of sodium hydroxide added (x-axis).

b   Use your graph to calculate the volume of sodium hydroxide added at the equivalence point for the titration. We will call this volume V.

c   Use your graph to find the pH when the volume of sodium hydroxide added is $\frac{1}{2} \times V$.

d   At $\frac{1}{2}V$, pH = pK_a. Use this information to calculate a K_a value for the acid.

e   A databook value for this same acid is given as 1.7×10^{-5} mol dm^{-3}. Suggest reasons for any difference between the sample data value and the databook value.

f   Suggest any modifications to the practical procedure that might improve the accuracy of your experiment.

Further questions and analysis

a   Another organic acid can be represented as HA(aq) and dissociates according to the equation: HA(aq) $\rightleftharpoons$ H$^+$(aq) + A$^-$(aq). Write an expression for K_a for this acid.

b   Using your answer to part **a**, explain why K_a = [H$^+$] at *half the volume of the equivalence point* when the acid is titrated with sodium hydroxide.

c   The pH at the equivalence point for the titration carried out above is about pH = 8.4. Using appropriate equations, explain why a pure solution of sodium ethanoate has an alkaline pH.

d   The ammonium ion in aqueous solution can act as an acid according to the following equation:

$$NH_4^+(aq) + H_2O(l) \rightleftharpoons NH_3(aq) + H_3O^+(aq)$$

The pK_a for this equilibrium is 9.3.

Use this information to calculate the pH of a $0.025\,mol\,dm^{-3}$ of $NH_4Cl(aq)$, stating any assumptions you have made.

Answers

Questions on data

a
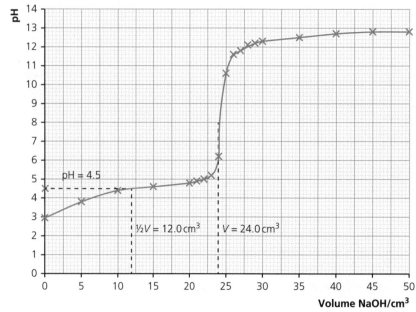

b   From the graph, extrapolate from the mid-point of the most vertical part of the titration curve down to the x-axis. $V = 24.0\,cm^3$

c $1/2\,V = 12.0\,cm^3$. Draw vertically up from the x-axis to the graph and then extend horizontally across until the y-axis is reached. pH = 4.5

d $pH = -\log_{10}[H^+]$. Thus $[H^+] = 10^{-4.5} = 3.16 \times 10^{-5}\,mol\,dm^{-3}$

e   Log scale is very 'unforgiving'. Small changes in the pH value result in big changes (by a factor of ten) in the non-log scale. So $-\log_{10}(1.7 \times 10^{-5}) = 4.77$, which is only 0.27 pH units away from the experimental value from the sample data. Also during the process of drawing the graph we have to estimate the mid-point of the vertical part of the curve and extrapolate.

f   The accuracy of the experiment depends on judging the correct equivalence point. If smaller volumes of alkali were added (e.g. $0.5\,cm^3$) as the equivalence point is approached, then a better approximation of the equivalence point could be made.

Practical tip

Make sure that you can correctly calculate inverse logs on your calculator. This may involve the shift function on your calculator, indicated as a 10^x symbol above the $\log_{10}$ function key.

Further questions and analysis

a $K_a = \dfrac{[H^+(aq)]\,[A^-(aq)]}{[HA(aq)]}$

b At exactly half the volume of the equivalence point, half of the acid has been neutralised. This means that we can assume there are the same number of deprotonated $A^-(aq)$ ions as HA molecules. If these values are equal, then they cancel in the equation above to give $K_a = [H^+]$.

c Sodium ethanoate is a salt of a weak acid and a strong base. It dissociates in water to form ethanoate ions and sodium ions. Sodium ions do not act as conjugate bases, but ethanoate ions are effective conjugate acids. Therefore the following equilibrium will establish:

$$CH_3COO^-(aq) + H_2O \rightleftharpoons CH_3COOH(aq) + OH^-(aq)$$

The forward reaction produces $OH^-(aq)$ ions and as $K_w = [H^+][OH^-]$ is a constant at 298 K, then the concentration of $H^+(aq)$ ions must drop to maintain K_w.

d A $0.025\,\mathrm{mol\,dm^{-3}}$ solution of NH_4Cl will be 0.025 with respect to $NH_4^+(aq)$ ions. $pK_a = 9.3$, so $K_a = 5.6 \times 10^{-10}\,\mathrm{mol\,dm^{-3}}$. For:

$$NH_4^+(aq) + H_2O(l) \rightleftharpoons NH_3(aq) + H_3O^+(aq)$$

the equilibrium lies far to the left-hand side. Then we can assume $[NH_4^+] = 0.025$. We will also that assume that $[NH_3] = [H_3O^+]$. Thus, simplifying $[H^+]^2 = K_a \times 0.025$. So $[H] = \sqrt{5.6 \times 10^{-10} \times 0.025} = \sqrt{1.4 \times 10^{-11}} = 3.74 \times 10^{-6}$. Taking the log and removing the negative, we get pH = 5.42.

We assume that the $Cl^-(aq)$ ion has a negligible effect as a conjugate base, that is, it will not accept H^+ ions to form molecular HCl(aq) in aqueous solution.

■ Core practical 10

Investigating some electrochemical cells

Introduction

An electrochemical cell is a device that uses a chemical redox reaction to generate a potential difference that can be used to drive a current. Many of the modern advances in portable communication devices are as much due to advances in electrochemical cells (more commonly referred to as batteries) as to developments in microchips.

In this practical you will construct a number of different electrochemical cells and measure the potential differences generated. You will also be asked to consider your knowledge of the metal reactivity series and how it relates to the potential differences measured.

Before we start with the practical it is worth making sure that we understand how electrochemical cell notation works. You will find, for example, that reversing the leads connected to the voltmeter changes the sign of the voltage displayed. Standard electrode potentials can also have positive or negative values, so there is the potential for confusion. Consider the electrochemical cell shown in Figure 11.

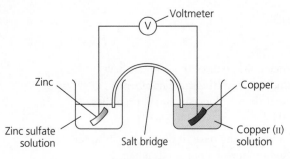

Figure 11

In abbreviated electrochemical cell notation this cell would be written as:

$$Zn(s)\,|\,Zn^{2+}(aq)\,\|\,Cu^{2+}(aq)\,|\,Cu(s)$$

In order to find a potential difference value for this cell you would look up the **standard electrode potential** for the two half-cells in your databook. The values given are −0.76 V for Zn and +0.34 V for Cu. To find the $E^{\ominus}_{cell}$ for this reaction we apply the following equation:

$$E^{\ominus}_{cell} = E^{\ominus}_{\text{right-hand half-cell}} - E^{\ominus}_{\text{left-hand half-cell}}$$

So for the above example:

$$E^{\ominus}_{cell} = +0.34 - (-0.76) = +1.10\,V$$

The electrons in this electrochemical cell flow from the Zn electrode to the Cu electrode. Thus the Zn must be connected to the negative terminal of the voltmeter and the copper to the positive. If you are reading a negative voltage, simply swap the leads around. Now you are ready for the practical.

Procedure

1 Clean the strips of zinc and copper using sandpaper.

2 Set up a zinc half-cell by pouring $50\,cm^3$ of zinc sulfate solution into the $100\,cm^3$ beaker and stand the cleaned strip of zinc in the beaker.

3 Repeat step 2, this time using copper sulfate solution and the cleaned strip of copper.

4 Prepare a salt bridge by soaking a folded piece of filter paper in a solution of potassium nitrate and place in the two beakers, as shown in Figure 13.

5 Connect the two metal strips to the voltmeter as shown in the diagram, making sure that the connecting clips are not in contact with the solutions.

6 Record the voltage displayed on the voltmeter to 2 d.p. If there is a negative symbol in front of the value, then swap the wire to the voltmeter.

7 Now repeat steps 1–6 for the combinations of metals and metal ion solutions shown in Table 14.

Practical tip

Note that the $\ominus$ symbol next to the E represents standard conditions. This means that the value is measured at 101 kPa pressure, 298 K and all solutions have a concentration of $1.0\,mol\,dm^{-3}$ with respect to the metal ions. Any conditions other than these will affect the value of the standard electrode potential.

Practical tip

Note that all solutions in this experiment are $1.0\,mol\,dm^{-3}$ with respect to the metal ion *except* the silver nitrate solution, which is $0.1\,mol\,dm^{-3}$.

Electrochemical cells to be tested with sample data

Table 14

Electrochemical cell	Voltage recorded
$Zn(s)\|Zn^{2+}(aq)\|\|Cu^{2+}(aq)\|Cu(s)$	+0.95
$Zn(s)\|Zn^{2+}(aq)\|\|Fe^{2+}(aq)\|Fe(s)$	+0.26
$Zn(s)\|Zn^{2+}(aq)\|\|Ag^+(aq)\|Ag(s)$	+1.34
$Fe(s)\|Fe^{2+}(aq)\|\|Cu^{2+}(aq)\|Cu(s)$	+0.66
$Cu(s)\|Cu^{2+}(aq)\|\|Ag^+(aq)\|Ag(s)$	+0.38

Questions on data

a **i** Given that the standard electrode potential for the copper half-cell is +0.34 V, calculate values for the Fe and Ag half-cells from the sample data above.

  ii Using your answer to part **a i**, calculate a value for the electrochemical cell shown below:

  $Fe(s)\|Fe^{2+}(aq)\|\|Ag^+(aq)\|Ag(s)$

b Using databook values for the standard electrode potentials of half-cells, calculations show that a value of +1.1 V is expected for the Zn/Cu electrochemical cell. Can you suggest why the value in the sample data is different?

Further questions

a **i** The databook value for the $Na(s)/Na^+(aq)$ half-cell is given as −2.71 V. Using this information, calculate the theoretical voltage generated by the $Na(s)\|Na^+(aq)\|\|Cu^{2+}(aq)\|Cu(s)$ electrochemical cell.

  ii Suggest why this cell would not actually generate this voltage in practice.

b The two half-cells in an electrochemical cell are connected by a salt bridge, as detailed in the practical procedure above. Explain in detail how the salt bridge connects the circuit. You should refer to the particles involved and how and why they move the way they do.

c An electrochemical cell is set up as shown in Figure 13. But before the circuit is connected a few drops of potassium manganate(VII) are put in the centre of the salt bridge between the two half-cells. Potassium manganate(VII) is a purple colour. Predict what you would observe when the cell is connected and explain your prediction.

Answers

Questions on data

a **i** Using $E^{\ominus}_{cell} = E^{\ominus}_{right\text{-}hand\ half\text{-}cell} - E^{\ominus}_{left\text{-}hand\ half\text{-}cell}$

  $Fe(s)\|Fe^{2+}(aq)\|\|Cu^{2+}(aq)\|Cu(s) = +0.66\ V$

Practical tip

Note that the ‖ symbol in the electrochemical cell notation represents the salt bridge. The salt bridge can be a piece of filter paper soaked in potassium nitrate. This solution is chosen as it is ionic and does not form precipitates with other ions.

$0.66 = 0.34 - x$, so $x = -0.32\,V$
So the half-cell for $Fe(s)/Fe^{2+}(aq) = -0.32\,V$ from sample data.

$Cu(s)|Cu^{2+}(aq)\|Ag^+(aq)|Ag(s) = +0.38\,V$

Using same equation as above, $0.38 = x - 0.34$, so $x = +0.72\,V$ for $Ag(s)/Ag^+(aq)$ half-cell.

ii We can now use these values to calculate a value for the $Fe(s)|Fe^{2+}(aq)\|Ag^+(aq)|Ag(s)$ cell.
$E^{\ominus}_{cell} = + 0.72 - (-0.32) = +1.04\,V$
Note that this value is based on the sample data and *not* the standard electrode potential data you will find in a databook.

b The measurements are unlikely to have been made under standard conditions (101 kPa and 298 K.)

Further questions

a i $Na(s)|Na^+(aq)\|Cu^{2+}(aq)|Cu(s)$
Using $E^{\ominus}_{cell} = 0.34 - (-2.71) = +3.05\,V$

ii This voltage will not be measured because the $Na(s)$ will react with the water in the $1.0\,M$ $Na^+(aq)$ solution:

$2Na(s) + 2H_2O(l) \rightarrow 2NaOH(aq) + H_2(g)$

b The salt bridge has $K^+(aq)$ and $NO_3^-(aq)$ ions in solution. Because zinc is more reactive than copper, the $Zn \rightleftharpoons Zn^{2+}(aq) + 2e^-$ lies to the right. Thus there is a net positive charge in the zinc sulfate solution (as there is a greater number of $Zn^{2+}(aq)$ ions). Therefore the $NO_3^-(aq)$ ions in the salt bridge move towards the zinc half-cell. The copper half-cell solution will have a *relative negative charge* and thus the $K^+(aq)$ ions will move toward the copper half-cell.

c Potassium manganate(VII) is $K^+(aq)$ and $MnO_4^-(aq)$ ions. The purple colour is due to the $MnO_4^-(aq)$ ions and thus once the cell is connected the circuit will be completed by the negative ions in the salt bridge moving towards the zinc half-cell (see above).

■Core practical 11

Redox titration

Introduction

You will be familiar with the technique of titrimetric analysis from CP2 and CP3. In CP11 the same technique is used, but is adapted to quantify the amount of iron (in the form of Fe^{2+} ions) per tablet. The beauty of this experiment is that no specific indicator is needed to signify the end point of the reaction, as the potassium manganate(VII) solution acts both as the oxidising agent and as the indicator.

Procedure

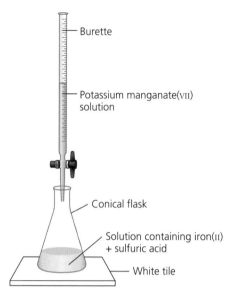

- Burette
- Potassium manganate(VII) solution
- Conical flask
- Solution containing iron(II) + sulfuric acid
- White tile

Figure 12

1 Crush five iron supplement tablets using a mortar and pestle.

2 Transfer the crushed tablet into a weighing boat and weigh the combined mass to at least 2 d.p. and record all masses in a suitable table.

3 Empty the contents of the weighing boat into a 250 cm^3 beaker and then reweigh the weighing boat.

4 Add 100 cm^3 of 1.5 mol dm^{-3} sulfuric acid to the beaker and stir to dissolve as much of the solid as possible.

5 Transfer the dissolved solution into a 250 cm^3 volumetric flask, making sure that no solid remains in the beaker. (You may add more sulfuric acid to the beaker to dissolve any remaining solid and transfer that to the volumetric flask.)

6 Make up the volumetric flask to a volume of 250 cm^3 using distilled water. You should stopper and invert your flask periodically as you make up the total volume to 250 cm^3.

7 Transfer about 50 cm^3 of solution from the volumetric flask into a clean dry beaker and then pipette 25.0 cm^3 of this solution into a 250 cm^3 conical flask.

8 Fill the burette with standardised 0.005 mol dm^{-3} KMnO$_4$(aq).

9 Titrate the contents of the conical flask with the KMnO$_4$(aq), swirling the contents of the flask as you add the standard solution (Figure 12).

10 The end point of the reaction is reached when a permanent faint pink colour remains for more than 30 seconds.

11 Repeat steps 7–10 until concordant titres are achieved, ensuring that all titres are recorded to the nearest 0.05 cm^3 in the results table.

Core Practicals

Sample results

Table 15

Mass of weighing boat + crushed tablets/g	2.75
Mass of empty weighing boat/g	0.89
Mass of tablets used/g	1.86

Table 16

	Rough	1	2	3
Volume of $KMnO_4$(aq) added/cm^3	20	19.80	19.70	19.75

Analysis of sample results

a Calculate a suitable mean titre from the results provided.

b Combine the two half-equations below to give a full ionic equation for the redox reaction taking place.

Oxidant: $8H^+(aq) + MnO_4^-(aq) + 5e^- \rightarrow Mn^{2+}(aq) + 4H_2O(l)$

Reductant: $Fe^{2+}(aq) \rightarrow Fe^{3+}(aq) + e^-$

c Calculate the number of moles of potassium manganate(VII) present in your mean titre.

d Use the information provided in the sample data together with your answer to part **c** to find the mass of iron per tablet.

e Complete Table 17 for the percentage error for each measurement made with each apparatus.

Table 17

Apparatus	Balance	Pipette	Burette
Error	±0.01 g	±0.05 cm^3	±0.05 × 2 cm^3
Percentage error from measurement	(0.01/1.86) × 100 = 0.54%	....	....

f Calculate the total percentage error for the experiment and then calculate the expected range for the mass of iron per tablet.

Further analysis questions

a The volumetric flask with the iron tablet solution was left un-stoppered overnight. When the experiment was repeated the next day with the same sample the amount of iron calculated per tablet dropped by 15%. Suggest why this might be the case.

b Explain why the potassium manganate(VII) acts both as an oxidant and as an indicator in this experiment.

c One of the filling agents used in iron tablets can be starch or cellulose. Can you suggest why after the end of the titration the end point colour may fade slowly after about 5 minutes? (Hint: the reaction mixture is acidic.)

Answers

Analysis of sample results

a $(19.80 + 19.70 + 19.75)/3 = 19.75 \, \text{cm}^3$

b $8H^+(aq) + MnO_4^-(aq) + 5Fe^{2+}(aq) \rightarrow 5Fe^{3+}(aq) + Mn^{2+}(aq) + 4H_2O(l)$

c Concentration = number of moles/volume in dm^3

Number of moles = $0.005 \times 19.75/1000 = 9.875 \times 10^{-5}$

d According to equation in **b**, $Fe^{2+}(aq):MnO_4^-(aq)$ ratio is 5:1, so the number of moles of $Fe^{2+}(aq)$ is $(5 \times 9.875 \times 10^{-5}) = 4.938 \times 10^{-4}$ moles in $25 \, \text{cm}^3$ of solution. In the total $250 \, \text{cm}^3$ there are 4.938×10^{-3} moles of Fe^{2+}. This has a mass of 55.8 (A_r of Fe) $\times 4.938 \times 10^{-3} = 0.276 \, \text{g}$ for the five tablets. Thus $0.276/5 = 0.055 \, \text{g}$ or $55 \, \text{mg}$ per tablet.

e

Apparatus	Balance	Pipette	Burette
Error	±0.01 g	±0.05 cm³	±0.05 × 2 cm³
Percentage error from measurement	(0.01/1.86) × 100 = 0.54%	(0.05/25) × 100 = 0.2%	(0.1/19.75) × 100 = 0.51%

f Thus the total possible error = $(0.54 + 0.2 + 0.51) = 1.25\%$

Thus the value range is $55 \pm (1.25/100 \times 55) = 55 \pm 0.69 \, \text{g}$ (range = 54.3–55.7 g) or 54–56 mg to 2 d.p.

Further analysis questions

a If the flask is left un-stoppered, oxygen in the air would oxidise Fe^{2+} ions to Fe^{3+} ions. Thus the titre value would be reduced by this number of Fe^{2+} ions oxidised.

b When manganate(VII) ions are reduced to Mn^{2+} ions, there is a colour change from purple to colourless. This colour change happens as long as there are Fe^{2+} ions present in solution to oxidise. As soon as the last Fe^{2+} is oxidised, the manganate(VII) ions stay in the oxidised form and their colour is so intense that the end point gives a definite pinkish tinge to the solution.

c The gradual fading of the pink colour suggests a further redox reaction. It is possible that, in the acid conditions, the starch is hydrolysed to glucose molecules and these can be further oxidised. (Glucose molecules have an aldehyde functional group, which can be oxidised to a carboxylic acid functional group.)

■ Core practical 12

The preparation of a transition metal complex

Introduction

A chemistry dictionary definition of a **complex** will say something like: 'A compound in which molecules or ions (in this context referred to as **ligands**) form coordinate bonds with a central metal atom or ion. Complex species maybe neutral or charged.'

Although complexes are covered in the inorganic section of an A-level chemistry courses, in the last 100 years or so it has become increasingly clear that an

A **ligand** is an ion or a molecule that donates an electron pair to a metal atom or ion to form a coordination complex.

understanding of complexes throws some light on how metal ions behave at the centre of enzymes and other metalloproteins. In many cases the metal at the centre of the biological complex is a transition metal and it is the transition metal's ability to shift between different oxidation states that is key to the functioning of the metalloprotein. The metalloprotein with which you will be most familiar is haemoglobin, a small section of which is shown in Figure 13.

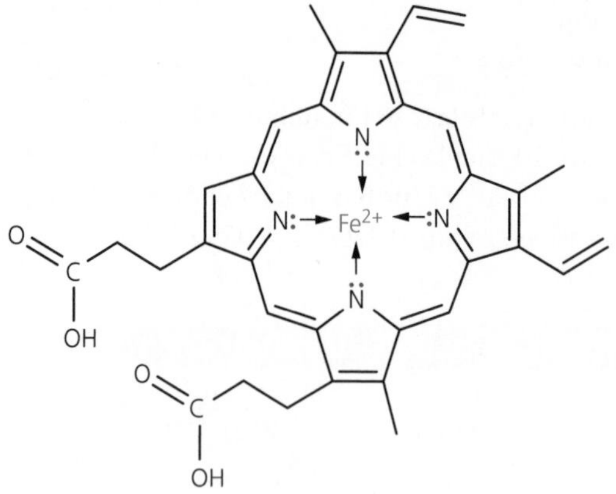

Figure 13

In this practical you will synthesise a copper(II) complex. Follow-up questions will ask you to calculate percentage yield and to think carefully about the types of ions or molecules that can behave as ligands.

Procedure

1 Weigh between 1.4 and 1.6 g of hydrated copper(II) sulfate. First weigh an empty test tube, add the copper sulfate and reweigh. The difference between the two masses is the mass of copper sulfate used, but be sure to record your mass of copper sulfate to 2 d.p.

2 Using a graduated pipette, add 4 cm³ of distilled water to the test tube.

3 Prepare a water bath using 100–150 cm³ of water in a 250 cm³ beaker and place the test tube in the beaker. Stir the solution to dissolve the copper sulfate.

4 Once dissolved, remove the test tube and take it to a fume cupboard.

5 Add 2 cm³ of concentrated ammonia to the copper sulfate solution while stirring to help dissolving. *It is important that this step is performed in a fume cupboard.*

6 Pipette 6 cm³ of ethanol into a 100 cm³ beaker and place in an ice bath.

7 Now pour the contents of the test tube into the ethanol and swirl the mixture.

8 Using a Büchner funnel and flask filter the contents of the 100 cm³ beaker and then rinse with a small volume of cold ethanol.

9 Pour the rinsings back through the Büchner funnel and, again, wash with a small volume of cold ethanol.

10 Scrape the crystals off the filter paper and pat dry using further sheets of filter paper.

11 Allow to dry in a warm place.

12 Now weigh the mass of the dry crystals, recording your value to 2 d.p.

Sample results

Table 18

Mass of test tube/g	19.75
Mass of test tube + hydrated copper(II) sulfate/g	21.23
Mass of hydrated copper(II) sulfate used/g	1.48
Final mass of complex crystals/g	1.08

Analysis of sample results

a i Calculate the number of moles of $CuSO_4.5H_2O$ used in the experiment.

ii Assuming that the formula of the crystallised complex is $Cu(NH_3)_4SO_4.H_2O$, calculate the number of moles of product collected.

iii Calculate the percentage yield in the experiment above.

iv Suggest *two* reasons why the experimental yield is different from the theoretical yield.

b Comment on the relative solubility of the complex in water and ethanol.

c Why is it important that the rinsing process is carried out with cold ethanol?

Further analysis questions

a When ammonia dissolves in water the following equilibrium is set up:

$$NH_3 + H_2O \rightleftharpoons NH_4^+ + OH^-$$

Of the four species in the above equation which one *cannot* behave as a ligand? Explain your answer.

b i Draw **skeletal formulae** for the following species:

- butan-1,4-dioate ion
- 1,2-diaminoethane

ii Explain how both species are able to behave as **bidentate ligands**.

iii Suggest why the hydrazine molecule (N_2H_4) *cannot* behave as a bidentate ligand.

c Look at the two amino acids shown in Figure 14.

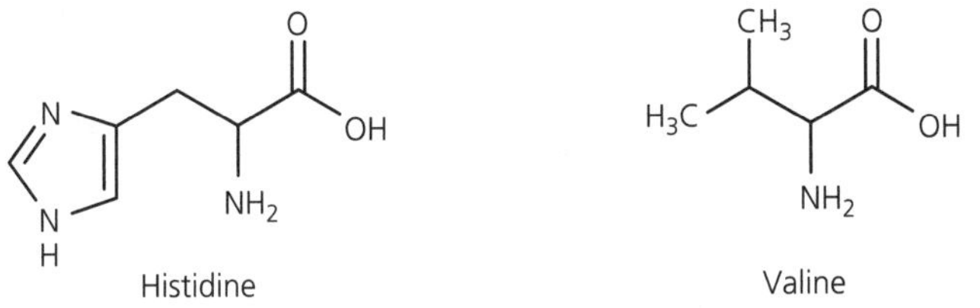

Figure 14

Suggest which of the two amino acids would be more likely to be found forming a coordinate bond with an Fe^{2+} ion at the centre of a metalloprotein. Explain your choice.

Answers

Analysis of sample results

a i $CuSO_4.5H_2O$ $M_r = 249.5\,g\,mol^{-1}$, so $1.48/249.5 = 5.93 \times 10^{-3}\,moles$

ii $Cu(NH_3)_4SO_4.H_2O$ $M_r = 245.5\,g\,mol^{-1}$, so $1.08/245.5 = 4.40 \times 10^{-3}\,moles$

Practical tip

Note that square brackets [ ] are only used when complexes have a charge such as $[CuCl_4]^{2-}$, but not if they are neutral such as $Fe(OH)_3(H_2O)_3$.

iii Yield $= (4.40 \times 10^{-3}/5.93 \times 10^{-3}) \times 100 = 74.2\%$

iv (1) Not all the product crystallised. (2) Some product was lost during the filtration process.

b The product is much less soluble in ethanol.

c The solubility of product is less in cold solvent than in hot solvent.

Answers to further analysis questions

a NH_4^+ cannot behave as a ligand as it is the only one of the four species not to have a **lone pair** available for the formation of a coordinate bond.

b i

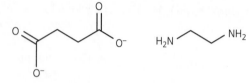

Butane-1,4-dioate ion   1,2-diaminoethane

> A **lone pair** is a pair of electrons in the outer shell of one of the atoms in a molecule or ion that is not involved in bonding.

ii A bidentate ligand forms two coordinate bonds per molecule. In the case of the butane-1,4-dioate ions, lone pairs on the O^- atoms form the coordinate bonds. In the case of the ethane-1,4-diamine, it is the lone pairs on the nitrogen atoms from the amine groups.

iii Hydrazine cannot form a bidentate ligand because, although it has two lone pairs available per anion, the coordinate bonds can only form at a 90° angle from each other, and because the lone pairs are at the end of a single covalent bond, it is structurally unable to form two bonds at 90° from each other.

c Histidine is able to form a coordinate bond due to a nitrogen on the ring part of the molecule. The side chain of valine is composed of three methyl groups and so is unable to form coordinate bonds.

■ Core practical 13a

Follow the rate of the iodine–propanone reaction using a titrimetric method

Introduction

The economic argument for investigating the speed or rate of a reaction is pretty clear; if I can make my product more quickly than my competitors then I stand to make more money. However, this is not the only reason to study rates of reaction. By looking at how changing the concentrations of different reactants affects the rate of a reaction, we can establish the **rate equation** for the reaction. This can then give insights into the **reaction mechanism** and the number and nature of the various steps involved in turning reactants into products.

Remember, the balanced chemical equation tells us only what we start and finish with. It tells us nothing about what happens in between. For this we need to look at the rate equation.

> A **reaction mechanism** shows how a reaction takes place in terms of the steps that involve bonds breaking and bonds forming. Mechanisms can involve heterolytic bonds forming and breaking (represented by a double-headed arrow), homolytic bonds (represented by a half-headed arrow) or ionic interactions such as the formation of a precipitate.

In this practical we will be using a continuous monitoring method that allows us to sample known volumes of the reaction mixture at given times, stop the reaction and then quantify the amounts of a particular reactant present. To determine the complete rate equation, the reaction would need to be repeated, so that the influence of each reactant (and the catalyst) could be quantified. In this particular experiment we will be investigating the effect of changing iodine concentrations on the initial rate of the reaction.

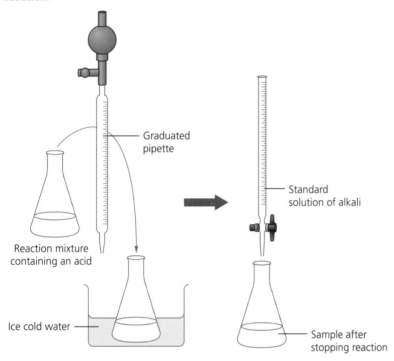

Figure 15

Procedure

1 Mix 25 cm³ of 1 mol dm⁻³ aqueous propanone with 25 cm³ of 1 mol dm⁻³ sulfuric acid in a 250 cm³ beaker.

2 Measure out 50 cm³ of 0.02 mol dm⁻³ iodine in potassium iodide solution in a measuring cylinder.

3 Add the iodine solution to the 250 cm³ beaker and immediately start the stopwatch.

4 After about 5 minutes withdraw 10 cm³ of the reaction mixture using a graduated pipette and add to a 100 cm³ conical flask (Figure 15).

5 The reaction in this 10 cm³ will continue until a spatula of sodium hydrogen carbonate is added to the 10 cm³ sample. *Make an exact note of the time that the sodium hydrogen carbonate is added to the sample.*

6 Repeat steps 4 and 5 for at least another five readings at about 5-minute intervals. *Note that you will have to refill your burette between readings.*

7 Each of the samples is now titrated against 0.01 mol dm⁻³ solution of sodium thiosulfate.

Core Practicals

8   Starting with the first sample taken at 5 minutes, add the sodium thiosulfate solution from the burette until the dark red/brown colour of the iodine solution fades to a yellow straw colour.

9   Add a few drops of starch indicator. A blue/black colour will be seen.

10 Now continue to titrate the solution with the sodium thiosulfate solution until the blue/back colour disappears.

11 Record the titre volume to ±0.05 cm³.

12 Repeat steps 8–11 for the remaining samples and record your results in a suitable table.

Sample results

Table 19

Time when sodium hydrogen carbonate added/min	5	11	15	21	25	31
Final burette reading/cm³	16.80	32.40	27.25	40.75	22.65	34.15
Initial burette reading/cm³	0.00	16.80	12.50	27.25	10.30	22.65
Volume of 0.01 mol dm⁻³ Na₂S₂O₃ added/cm³	16.80	15.60	14.75	13.50	12.35	11.50

Analysis of sample results

a   Plot the sample results from Table 19 with time in minutes on the x-axis and volume of sodium thiosulfate in cm³ on the y-axis.

b   Write a balanced full ionic equation for the reaction between sodium thiosulfate and iodine in solution.

c   With reference to your graph in part **a**, determine the **order of reaction** with respect to iodine. Explain your answer

d   In determining the order of reaction with respect to iodine, both the propanone and the acid are in a large excess. Explain why this needs to be the case.

e   Explain the role of sodium hydrogen carbonate in stopping the reaction. Write an ionic equation to help explain your answer.

Further analysis questions

a   Further investigation into this reaction shows that the full rate equation for the reaction is:

rate = $k[(CH_3)_2CO(aq)][H^+(aq)]$

i   What would be the units of k in this reaction?

ii   What effect would a *doubling* of *both* species in the rate equation have on the initial rate of this reaction?

iii How could you show from a concentration versus time graph that the order of reaction with respect to propanone is 1?

Practical tip

Note that the starch indicator is not added at the start because if the concentration of the iodine is too high, then the starch indicator will form an insoluble substance that will not dissolve easily at the end point.

The **order of reaction** with respect to a particular reactant is the power to which the concentration of a reactant is raised in the rate equation. At A-level you will only come across values of 0, 1 or 2. The total order of reaction is the sum of the powers in the full rate equation.

b The stoichiometric equation for this reaction can be shown as follows:

$$(CH_3)_2CO(aq) + I_2(aq) + H^+(aq) \rightarrow CH_3COCH_2I + 2H^+(aq) + I^-(aq)$$

  i What is the role of the H^+(aq) in this reaction? Explain your answer.

  ii How can we tell from the rate equation that the reaction mechanism *must* involve at least two steps?

  iii Using your knowledge of organic chemistry, propose a mechanism that is consistent with the rate equation shown. You should identify the rate-determining step.

Answers

Analysis of sample results

a

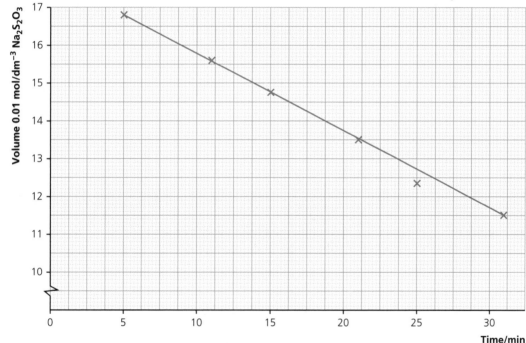

b $2S_2O_3^{2-}$(aq) + I_2(aq) $\rightarrow$ $S_4O_6^{2-}$ + $2I^-$(aq)

c Order with respect to iodine = 0. The rate of reaction is a constant (shown by the constant gradient) irrespective of iodine concentration (which is proportional to titre).

d Propanone and acid need to be in excess so that their concentrations remain effectively constant as the reaction progresses. If this were not the case, the change in rate could be due to the changing iodine concentration *or* the changing concentration of one of the other reactants.

e Sodium hydrogen carbonate neutralises the acid and thus removes one of the reactants:

$$HCO_3^-(aq) + H^+(aq) \rightarrow H_2O(l) + CO_2(g)$$

Further analysis questions

a i Making k the subject, $k = \text{rate}/[(CH_3)_2CO(aq)][H^+(aq)]$. Inputting units $\text{mol}\,dm^{-3}\,s^{-1}/\text{mol}^2\,dm^{-6}$ and cancelling units $= dm^3\,mol^{-1}\,s^{-1}$.

ii Doubling each individually would double the reaction rate, so doubling both will give a ×4 rate increase.

iii A first-order reaction profile would show an exponential decay *with a constant half-life*. Measure how long the initial concentration takes to become half. Then measure how long this half concentration takes to become one quarter of the original concentration. If the time values for these two changes are the same, then the reaction is first order with respect to that reactant (Figure 16).

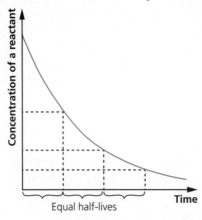

Figure 16

b i The $H^+(aq)$ is behaving as a **catalyst**. Note that it is regenerated at the end of the reaction.

ii If the rate equation does not have I_2 in it, but the final product does have an I atom, then there must be at least one further step that does involve the I_2 molecule.

iii

The suggested mechanism has a slow or rate-determining step that leads to the formation of an intermediate (propen-2-ol). The second, fast step leads to the formation of the product.

■ Core practical 13b

Investigating a 'clock reaction' to determine a rate equation

Introduction

In this practical we are going to investigate the reaction between iodide ions and peroxodisulfate ions in an attempt to determine the full rate equation for the reaction.

The full ionic reaction is given by the following equation:

$$S_2O_8^{2-}(aq) + 2I^-(aq) \rightarrow 2SO_4^{2-}(aq) + I_2(aq)$$

It is worth restating that this stoichiometric reaction tells us nothing about the reaction mechanism. We can appreciate that the likelihood of this reaction taking place in a single step is *extremely* unlikely (consider the probability of a simultaneous collision between two iodide ions and one peroxodisulfate ion, in only those proportions with sufficient energy in solution). Thus we have to conclude that the reaction must take place in a number of steps. Determining the rate equation for this reaction will help us gain an insight into the nature of those steps.

Measuring the rate of iodine formation in the reaction above is difficult as the formation of iodine leads to a *gradual* change in the colour of the solution from colourless to red/brown. This means no clear end point can be seen. The problem is overcome by adding a fixed number of moles of sodium thiosulfate together with a few drops of a 1% starch solution. Any iodine formed in the reaction is immediately converted back into iodide according to the following equation:

$$I_2(aq) + 2S_2O_3^{2-}(aq) \rightarrow 2I^-(aq) + S_4O_6^{2-}(aq)$$

Once this fixed amount has been used up, the molecular iodine produced in the first reaction stays in solution long enough to react with the starch indicator and give a clear and immediate blue/black colour. This provides a clear end point for the reaction.

Procedure

1 Measure out $10\,cm^3$ of $0.1\,mol\,dm^{-3}$ KI(aq) into a $100\,cm^3$ conical flask and place onto a white tile.

2 Add $5\,cm^3$ of $0.025\,mol\,dm^{-3}$ sodium thiosulfate solution to the conical flask. followed by five to six drops of freshly prepared 1% starch indicator solution.

3 Measure out $10\,cm^3$ of $0.1\,mol\,dm^{-3}$ sodium peroxodisulfate solution into a measuring cylinder.

4 Start the reaction by adding the contents of the measuring cylinder to the conical flask and immediately start the stopwatch.

5 Record the time taken for the blue/black colour to appear to the nearest second.

6 Repeat steps 1–6 for the volumes of the reactants detailed in the sample results table (Table 20).

Core Practicals

Sample results

Table 20

Experiment number	Volume of KI(aq)/cm³	Volume of Na₂S₂O₈(aq)/cm³	Volume of Na₂S₂O₃(aq)/cm³	Volume of water added/cm³	Time taken/s
1	10	10	5	0	45
2	10	8	5	2	56
3	10	6	5	4	77
4	10	4	5	6	110
5	10	2	5	8	238
6	8	10	5	2	58
7	6	10	5	4	78
8	4	10	5	6	107
9	2	10	5	8	242

All reactions carried out at 298 K

Processing and analysis of sample results

a Complete Table 21 using the results shown in Table 20.

Table 21

Experiment number	Concentration of KI(aq)/mol dm⁻³	Concentration of Na₂S₂O₈(aq)/mol dm⁻³	Rate/10⁻² s⁻¹
1	0.040	0.040	2.2
2			
3			
4			
5			
6			
7			
8			
9			

b Plot the following two sets of data on a suitable graph:
  i   Concentration of potassium iodide in mol dm⁻³ versus reaction rate in 10⁻² s⁻¹.
  ii Concentration of sodium peroxodisulfate in mol dm⁻³ versus reaction rate in 10⁻² s⁻¹.

c Use your results from part **b** to calculate the orders of reaction with respect to iodide and peroxodisulfate and determine the rate equation.

d Calculate a value for the rate constant in this reaction, making sure you include units in your answer.

e Use your answers above to predict which species are involved in the rate-determining step of the reaction.

f Further investigations into this reaction show that the presence of Fe^{3+}(aq) ions catalyses this reaction. Suggest what would be the effect on:
  i   the value of the rate constant k
  ii the rate equation

Practical tip

You will need to calculate the total volume of the solutions used in each reaction in order to calculate the concentration of each reactant.

Practical tip

By multiplying your $1/t$ value by 100 you will get an easier number to plot but don't forget to divide this number by 100 when calculating a value for rate constant k.

Answers

Processing and analysis of sample results

a

Experiment number	Concentration of KI(aq)/mol dm^{-3}	Concentration of Na$_2$S$_2$O$_8$(aq)/mol dm^{-3}	Rate/10^{-2} s^{-1}
1	0.040	0.040	2.2
2	0.040	0.032	1.8
3	0.040	0.024	1.3
4	0.040	0.016	0.91
5	0.040	0.008	0.42
6	0.032	0.040	1.7
7	0.024	0.040	1.3
8	0.016	0.040	0.93
9	0.008	0.040	0.41

b i

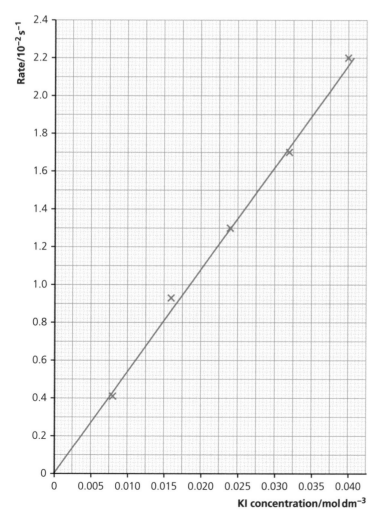

ii

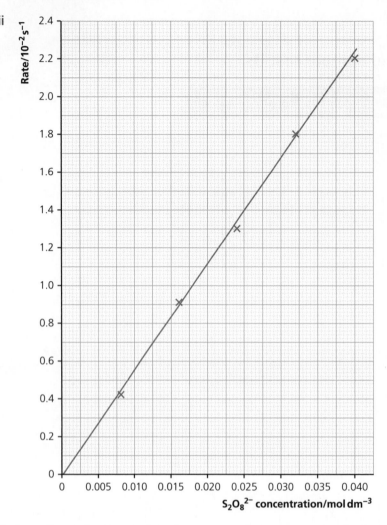

Rate/$10^{-2}\,\mathrm{s}^{-1}$ vs $S_2O_8^{2-}$ concentration/mol dm^{-3}

c Both are first order. Rate is directly proportional to concentration for both reactants (gradient is constant). Thus rate = $k[S_2O_8^{2-}][I^-]$.

d Using experiment 1, k = rate/$[S_2O_8^{2-}][I^-]$ = 0.022/(0.04 × 0.04) = 13.75 dm^3 mol^{-1} s^{-1}.

e Rate determining step must be the first step involving:

$S_2O_8^{2-} + I^- \rightarrow$ intermediate

f i The value of k would be larger. By lowering the activation energy, more reaction collisions would result in a reaction and thus the reaction rate for the same concentrations of reactants would be greater.

ii The rate equation would be different. The catalyst must provide an alternative pathway with a different reaction mechanism. If the catalyst lowers the activation energy, the rate-determining step of this new reaction mechanism must involve the catalyst. Thus the catalyst would appear in the rate equation.

Core practical 14

Finding the activation energy of a reaction

Introduction

Students carrying out reactions in the laboratory rarely stop to consider the importance of the energy required to initiate a reaction, partly because the laboratory scale of the reaction is so small. On an industrial scale this is a different matter. In many industrial chemistry processes the energy requirements are one of the most significant costs to be considered and will thus be reflected in the final cost of the product. For this reason it is important to be able to determine the **activation energy** of a reaction. Once established, synthetic routes using different catalysts can be compared to see which method is the most cost efficient.

We have considered the general form of the rate equation in CP13. For a reaction A + B → products, the rate equation has the form:

rate = $k[A]^m[B]^n$

where m and n are experimentally determined values (the orders of reaction with respect to A and B) and k is a temperature-dependent rate constant.

There is a mathematical relationship, known as the Arrhenius equation, between the temperature-dependent rate constant and the activation energy for a reaction, which can be summarised as follows:

$k = Ae^{-E_a/RT}$

or using natural logarithms:

$\ln k = \ln A - \dfrac{E_a}{RT}$

This relationship is useful because we can see that k is related to the activation energy of the reaction. If we have established the order values in the rate equation, we can determine a value for k. However, k is a **temperature-dependent constant** and thus will change with temperature T. By investigating how k varies with T, we can obtain a value for E_a.

In this experiment we can adapt the Arrhenius equation in such a way that we do not need to know the full rate equation for the reaction. This will be discussed in more detail after the sample results.

The **activation energy** for a reaction is represented by the symbol (E_a) and is defined as the minimum energy required for a chemical reaction to take place.

Practical tip

It is easy to get confused between 'k' the rate constant and 'K' the equilibrium constant. The equilibrium constant will always be a capital K and should have a subscript c, p a or w, as required. The rate constant is a small k and will never have a subscript letter.

Procedure

1 Pipette $10\,cm^3$ of phenol solution and $10\,cm^3$ of prepared bromide/bromate(v) solution into a boiling tube.

2 Add four to five drops of methyl red indicator to the mixture.

3 Pipette $5\,cm^3$ of sulfuric acid solution into another boiling tube.

4 Use a kettle to prepare a beaker of water to a temperature of $75°C \pm 1°C$ and place both boiling tubes into the beaker.

5 Allow sufficient time for both solutions to equilibrate with the temperature of the water in the beaker and then add the contents of the two boiling tubes together and start the stopwatch.

6 Leave the reaction boiling tube in the water and record the time it takes for the red colour of the methyl indicator to turn colourless.

7 Repeat steps 1–6 for six more temperatures of approximately 65, 55, 45, 35, 25 and 15°C. (Note it does not matter if these temperatures are exactly as stated as long as you do record the temperature to the nearest 1°C.) You will need ice to cool the water to below 20°C.

8 Record all your results in the table provided.

Sample results

Table 22

Temp/°C	Time/s	Temp/K	$\frac{1}{T} \times 10^{-3}/K^{-1}$	ln t
16	430	289	3.46	6.1
25	220	298	3.36	5.4
34	96	307	3.26	4.6
44	47	317	3.15	3.9
56	19	329	3.04	2.9
65	10	338	2.96	2.3
91	3	364	2.75	1.1

Analysis of sample results

A modification of the Arrhenius equation is used in the analysis of this reaction. The rate constant k is proportional to time t: $k \propto 1/t$ or $k = c/t$, where c is a constant. Substituting c/t for k in the Arrhenius equation we get:

$$\ln c/t = \ln A - \frac{E_a}{RT}$$

Rearranging this equation we get:

$$\ln c - \ln t = \ln A - \frac{E_a}{RT}$$

and finally:

$$\ln t = (\ln c - \ln A) + \frac{E_a}{RT}$$

This equation fits the general formula of a straight-line equation $y = mx + c$. If we plot $\ln t$ against $1/T$, then the gradient of the line will equal $+E_a/R$. (Note that this is different from a plot of $\ln k$ versus $1/T$, which gives a gradient of $-E_a/R$.)

a **i** Plot a graph of the sample results with $\ln t$ on the y-axis and $1/T \times 10^{-3}\,K^{-1}$ on the x-axis and connect the points with a line of best fit.

  ii Calculate the gradient of your line. (Remember that the x-axis number must be multiplied by 10^{-3}.)

  iii Given that the gradient of the graph $= E_a/R$, where $R = 8.31\,Jmol^{-1}\,K^{-1}$, calculate a value for E_a, giving your answer in $kJ\,mol^{-1}$.

  iv Which species is responsible for decolorising the methyl red indicator at time t in each experiment? Classify the type of reaction.

  v Explain the function of the phenol in this reaction.

Further analysis questions

a A full analysis of this reaction generates the following rate equation:

  rate $= k[BrO_3^-][Br^-][H^+]^2$

  i What is the total order for this reaction?

  ii What are the units for k in this reaction?

  A suggested mechanism for this reaction in five steps is as follows:

  Step 1: $H^+(aq) + BrO_3^-(aq) \rightleftharpoons HBrO_3(aq)$

  Step 2: $H^+(aq) + HBrO_3(aq) \rightleftharpoons H_2BrO_3^+(aq)$

  Step 3: $H_2BrO_3^+(aq) + HBr(aq) \rightarrow HBrO(aq) + HBrO_2(aq)$

  Step 4: $HBrO_2(aq) + HBr(aq) \rightarrow 2HBrO(aq)$

  Step 5: $HBrO(aq) + HBr(aq) \rightarrow Br_2(aq) + H_2O(l)$

  iii Combine the five steps to generate the overall stoichiometric equation for the reaction.

  iv Using the rate equation in part **a**, suggest which of the steps is the rate-determining step, giving a reason for your choice.

Practical tip

Remember: stoichiometric means the correct numbers in front of reactants and products to give the full balanced equation.

Answers

Analysis of sample results

a i

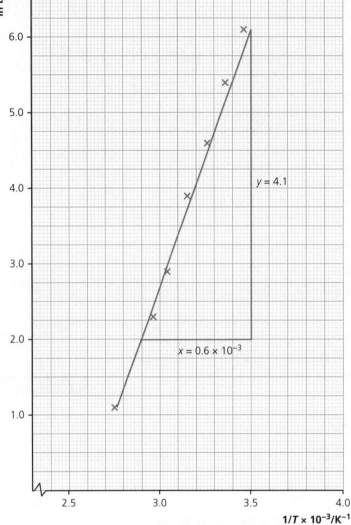

ii From graph: $y/x = 4.1/0.6 \times 10^{-3} = 6833.3 = E_a/R$, which is the gradient.

iii Taking R as 8.31 J mol^{-1} K^{-1}, then $E_a = 56\,785$ J or 57 kJ mol^{-1} to 2 s.f.

iv Molecular bromine (Br_2) decolorises the methyl red. It is acting as an oxidising agent.

v The bromine combines with the phenol before it has a chance to decolorise the methyl red. The fixed quantity of phenol in each reaction means that the end point of the reaction is always for the same amount of bromine produced.

Further analysis questions

a i Total order of reaction = 4. (The sum of all the powers to which the reactant species in the rate equation are raised.)

ii $k = \text{rate}/(\text{mol dm}^{-3})^4 = \text{dm}^9 \text{mol}^{-3} \text{s}^{-1}$

iii $6H^+(aq) + BrO_3^-(aq) + 5Br^-(aq) \rightarrow 3Br_2 + 3H_2O(l)$

iv Step 3 is the rate-determining step. The two fast previous steps each involve H^+ ions, hence the rate-determining step is dependent on the completion of these two steps *and* step 3.

■ Core practical 15

Analysis of some inorganic and organic unknowns

Introduction

This is the second practical that requires you to carry out some qualitative tests on unknowns. As with practical CP7, you will be asked to carry out tests that enable you to identify both anion and cation in the inorganic unknowns and to be able to identify the functional group for the organic unknowns. There are four inorganic unknowns labelled A–D, each of which will have one anion and one cation. The three organic unknowns will have one functional group. Rather than provide you with a list of instructions to follow, the emphasis in this practical will be on you carrying out the necessary research into the tests before tackling the practical. You should consider how to be systematic and logical in your approach and think about the order in which you carry out the tests. You should also pay particular attention to issues of safety.

Tests to be carried out on unknowns A–D, and sample results

Table 23

Test	A	B	C	D
Flame test	No colour observed	Lilac colour	Bright yellow/ orange colour	Blue/green flame
Test with NaOH(aq)	Grey green precipitate	No reaction	No reaction	Blue precipitate
Test for halide ion	Cream precipitate formed	No reaction	No reaction	White precipitate
Test for sulfate ion	No precipitate formed	No reaction	White precipitate	No reaction
Test for carbonate ion	No reaction	Effervescence; gas turned lime water cloudy	No reaction	No reaction

Practical tip

You should to be able to give a clear account of how to carry out a flame test.

Tests to be carried out on unknowns X, Y and Z, and sample results

Table 24

Test	X	Y	Z
Alkene functional group	No reaction	No reaction	Bromine water turned from orange to colourless
Aldehyde functional group	Orange precipitate	No reaction	No reaction
Carboxylic acid functional group	No reaction	Effervescence turned lime water cloudy	No reaction

Analysis of sample results

a Identify unknowns A–D from the data in Table 23.

b Unknown A gave a grey/green precipitate with NaOH(aq).

  i Give the full chemical formula for the species responsible for the precipitate.

  ii Describe and explain what you would expect to see if this test were left to stand for 5 minutes.

c Give details of how you would test for the presence of a halide ion, including how you would eliminate the chance of a false positive.

d Both unknowns A and D formed a precipitate with silver nitrate. Give details of a further test that would enable you to distinguish the two halide ions.

e Substance B reacted with lime water to turn the solution cloudy. Give a full equation for this reaction and name the substance responsible for the cloudiness.

f Unknown Z gave a positive result for the test with bromine water.

  i Give details of any precautions that need to be taken when carrying out this test, explaining your reasons.

  ii Assuming that substance Z is cyclohexene, give the formula for the product and name it.

  iii Could there be more than one isomer from the reaction between bromine water and cyclohexene? Explain your answer.

g Give details of the test for the aldehyde functional group and name the substance responsible for the orange precipitate formed with unknown X.

h Give details of the test for a carboxylic acid functional group that would account for the result given with unknown Y.

i It was found that a sample of X left exposed to air for a period of time gave a slight effervescence when the test for the carboxylic acid was carried out. Suggest an explanation for this observation other than contamination.

Answers

Analysis of sample results

a $A = FeBr_2$; $B = K_2CO_3$; $C = Na_2SO_4$; $D = CuCl_2$

b i $Fe(OH)_2(H_2O)_4$

ii After standing in air the precipitate would turn a rusty brown colour at the top. This is due to the Fe^{2+} being oxidised by oxygen in the air to $Fe(OH)_3(H_2O)_3$ (containing the Fe^{3+} ion).

c A sample of the solid is dissolved in a small amount of distilled water. Drops of nitric acid are added and then drops of silver nitrate added. The precipitate formed is the silver halide. Nitric acid is added to remove any dissolved carbonate ions that can form due to carbon dioxide from the air dissolving. The false positive would be due to the formation of insoluble silver carbonate.

d The difference between the colours is difficult to see, so a further test would be to see whether the precipitate dissolves in aqueous ammonia; silver chloride will dissolve whereas silver bromide will not.

e $CO_2(g) + Ca(OH)_2(aq) \rightarrow CaCO_3(s) + H_2O(l)$

Cloudiness is due to insoluble calcium carbonate.

f i Test should be carried out in the absence of sunlight. Otherwise decolorisation will be due to free radical mechanism.

ii $C_6H_{12}Br_2$; 1,2-dibromocyclohexane

iii Due to the lack of free rotation around the C–C bond in the ring, the bromines may be on opposite sides of the ring (*trans* or *E* isomers) or the same side (*cis* or Z isomers).

(Note that addition of OH^- in the second part of the electrophilic addition can lead to 2-bromocyclohexanol products.)

g The test is to use Fehling's (solutions 1 and 2). The orange precipitate is copper(I) oxide formed from the reduction of Cu^{2+} ions in Fehling's solution 2. The aldehyde functional group is oxidised to a carboxylate group.

h A sample of solid sodium carbonate is added to the test substance. The presence of an acid group leads to the production of $CO_2(g)$, which can be tested for by bubbling through lime water.

i As X has the aldehyde functional group, if left exposed to air it can oxidise to a carboxylic acid group. This would account for the slight effervescence.

X = propanal; Y = ethanoic acid; Z = cyclohexene

■ Core practical 16

The preparation of aspirin

Introduction

The final practical involves the synthesis of aspirin. Aspirin is still one of the most widely taken medicines throughout the world and has been used in some form since 3000 BCE. It can be used for pain relief, as an anti-inflammatory and to reduce fever. More recently it has been prescribed as an anticoagulant to help prevent coronary thrombosis for those considered at high risk.

There are a number of ways of synthesising aspirin, but in this procedure the starting materials are ethanoic anhydride and 2-hydroxybenzoic acid (Figure 17).

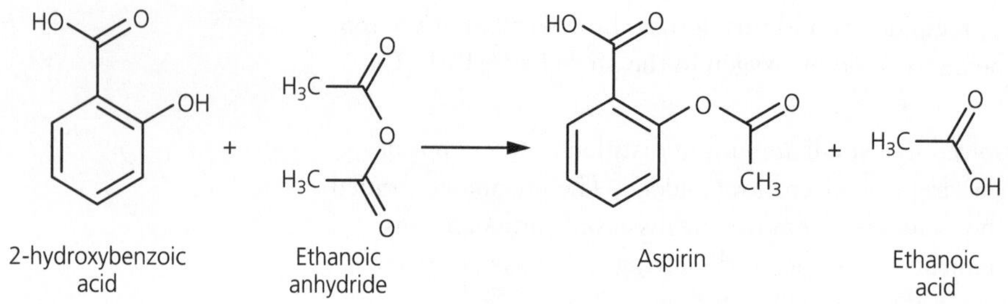

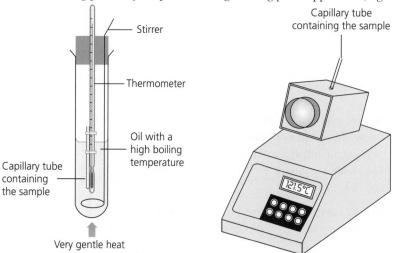

2-hydroxybenzoic acid + Ethanoic anhydride → Aspirin + Ethanoic acid

Figure 17

Procedure

1 Weigh approximately 2 g of 2-hydroxybenzoic acid into a weighing boat and record its mass to 2 d.p.

2 Transfer the 2-hydroxybenzoic acid into a pear-shaped flask and reweigh the weighing boat.

3 Clamp the pear-shaped flask so that it is suspended in a beaker of cold water.

4 Add 5 cm³ of ethanoic anhydride to the pear-shaped flask, followed by five drops of concentrated sulfuric acid, and swirl the mixture so that the solid dissolves.

5 Fix a condenser on the flask in such a way that you can reflux the mixture.

6 Using a Bunsen burner heat the beaker gently for about 10 minutes.

7 Remove the pear-shaped flask from the beaker and add some crushed ice and some distilled water.

8 Place the pear-shaped flask in an ice bath until the product precipitates out.

9 Filter off the solid product using a Büchner funnel and suction apparatus.

10 Wash the crude product with a small volume of ice-cold water.

11 Recrystallise the raw product using the minimum volume of hot solvent made from a 1:3 volume ratio of ethanol and water.

12 Filter the recrystallised product and dry.

13 Measure the mass of your product to the nearest 2 d.p.

14 Check the melting point of your product using melting point apparatus (Figure 18).

Stirrer

Thermometer

Oil with a high boiling temperature

Capillary tube containing the sample

Very gentle heat

Capillary tube containing the sample

121.5°C

Figure 18

Sample results

Table 25

	Mass/g	Melting point/°C
Weighing boat + 2-hydroxybenzoic acid	3.12	
Empty weighing boat	1.09	
Mass of 2-hydroxybenzoic acid	2.03	
Mass of pure product	2.06	130–132

Analysis of sample results

a **i** Using Figure 17 calculate the molar masses of 2-hydroxybenzoic acid and aspirin.

   ii Calculate the atom economy for this reaction.

   iii Calculate the percentage yield.

   iv Suggest *two* reasons for any loss in product.

b In step 7 in the reaction procedure water is added to break down any unreacted ethanoic anhydride. Write an equation for this reaction.

c Ethanoic anhydride has a density of $1.08\,g\,cm^{-3}$. Show that the amount of ethanoic acid used in this experiment is in excess.

Further analysis questions

a An alternative synthetic method uses ethanoyl chloride in place of ethanoic anhydride.

   i Write a balanced equation for this reaction.

   ii Calculate the atom economy for this second process.

   iii What other considerations need to be taken into account when choosing either of the synthetic routes for an industrial-scale synthesis?

b Old aspirin left in a moist environment can smell like vinegar. Explain why this is the case.

c There are three possible isomers of hydroxybenzoic acid. Draw all three and suggest which instrumental analytical technique you could use to distinguish between the three isomers.

Answers

Analysis of sample results

a **i** 2-hydroxybenzoic acid $138\,g\,mol^{-1}$; aspirin $180\,g\,mol^{-1}$

   ii Molar mass of reactants = 138 + 102 = 240. Molar mass of aspirin = 180. Atom economy = $(180/240) \times 100 = 75\%$.

   iii Moles of 2-hydroxybenzoic acid = 2.03/138 = 0.0147. Theoretical 100% yield means $0.0147 \times 180 = 2.65\,g$. Actual yield = 2.06, so percentage yield = $(2.06/2.65) \times 100 = 78\%$.

   iv (1) Reaction might not have gone to completion. (2) Solid lost during filtration.

b $(CH_3CO)_2O + H_2O \rightarrow 2CH_3COOH$

Core Practicals

c Mass = density × volume. 1.08 × 5 = 5.4 g. Molar mass of ethanoic anhydride = 102. Moles of ethanoic anhydride = 5.4/102 = 0.053 moles. The reaction ratio is 1:1, so this is well in excess of 0.0147 moles of 2-hydroxybenzoic acid.

Further analysis questions

a i

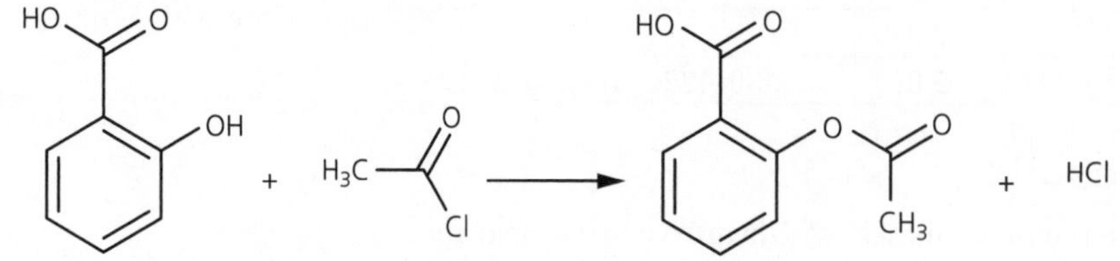

ii Total mass of reactants = 216.5. Atom economy = (180/216.5) × 100 = 83%.

iii Even though the second method has a higher atom economy the by-product is HCl, which is a strong acid. This has implications for disposal and the environment, and may also corrode apparatus.

b The ester functional group will slowly hydrolyse, giving 2-hydroxybenzoic acid and ethanoic acid, which smells of vinegar.

c

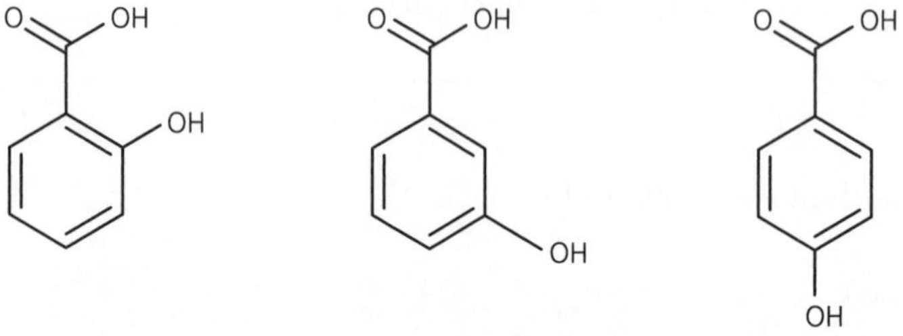

2-hydroxybenzoic acid   3-hydroxybenzoic acid   4-hydroxybenzoic acid

^{1}H NMR would enable you to distinguish the different isomers, focusing in on the type and number of aromatic proton environments and their shifts.

Questions & Answers

This Questions & Answers section contains 23 exam-style questions with mark allocations and comments on how to answer them. Questions marked with an *indicate AS material and questions based on practical chemistry that might appear on papers 1 and 2 at AS or papers 1 and 2 at A-level. The remaining questions are based on A-level material only and could be encountered on any of the three papers at A-level. *Note that papers 1 and 2 for the AS are not identical to papers 1 and 2 in the full A-level.* As a rough guide, the time allowance for each question is 1 mark per minute, but this is *only* a guide. Note that in some of the extended questions there will be more possible scoring points than marks available for the question. So, for example, in a 12-mark question there may be 15 possible ways of scoring the maximum 12 marks.

A formulae and data sheet is provided with each test. Copies may be downloaded from the Edexcel website, or can be found at the end of past papers.

In addition to the written papers, there is a *Science Practical Endorsement*, for which you will be assessed separately. The Endorsement will not contribute to the overall grade for your A-level qualification, but the result will be recorded on your certificate. To gain the Endorsement, you need to provide evidence that you have completed the necessary practical activities as indicated in the Edexcel specification. You must complete a minimum of 12 identified practical activities that demonstrate your competence to:

- follow written procedures
- apply investigative approaches and methods when using instruments and equipment
- safely use a range of practical equipment and materials
- make and record observations
- research, reference and report

You should refer to the Edexcel specification, where there is a table in Appendix 5d showing how each core practical activity maps to the required practical techniques.

Examiners use certain command terms that require you to respond in a particular way: for example, 'state', 'explain' or 'discuss'. You must be able to distinguish between these terms and understand exactly what each requires you to do. A full list of command terms can be found in the Edexcel specification, Appendix 7. You are strongly recommended to download this (www.edexcel.com) and print-off a copy for reference.

You should pay particular attention to diagrams, drawing graphs and making calculations. Many students lose marks by failing to label diagrams properly, not giving essential data on graphs and, in calculations, by not showing all the working or by omitting units.

In this section, each question is followed by a brief analysis of what to watch out for when answering the question, ⓔ. Student responses are followed by a comment, preceded by the icon ⓔ, indicating where credit is due.

■ Practical exam-style questions

Question 1*

Determining the concentration of sodium hydroxide by titration.

A student is asked to determine the concentration to 3 s.f. of a sodium hydroxide solution that is approximately $0.1\,mol\,dm^{-3}$ by titrimetric analysis. There is a choice of two possible standardising reagents that can be used: (i) $250\,cm^3$ of $1.00\,mol\,dm^{-3}$ solution of $HCl(aq)$ and (ii) $100\,g$ of sulfamic acid (NH_2SO_3H).

(a) Give details of how you would prepare a suitable standardising solution from each of the two reagents above. You should include suitable quantities of each regent and a clear method of how each standard is prepared. (6 marks)

ⓔ As with all questions, it is important to look at the number of marks available. As a rough guide, the examiner will be looking for 3 marks per standardising agent.

It is worth remembering that titrations should never be carried out using concentrated solutions. Generally, solutions of no greater concentration than $0.1\,mol\,dm^{-3}$ should be used.

Student answer

(a) (i) For $HCl(aq)$ the standard must be diluted to a value of $0.100\,mol\,dm^{-2}$ ✓. This should be done using suitable volumetric apparatus, e.g. a $25\,cm^3$ volumetric pipette should be used to remove the HCl *and* this should be added to a $250\,cm^3$ volumetric flask ✓. The total volume should be made up to $250\,cm^3$ using distilled water ✓ⓐ, so that the meniscus sits on the calibration mark ✓ⓐ.

ⓔ ⓐ Alternatives for this mark.

(ii) For the sulfamic acid a suitable mass of the acid should be chosen to make a standard of approximately $0.1\,mol\,dm^{-3}$ (but value must be to 3 d.p.) ✓. Suitable mass should be shown by calculation, e.g. 0.025 mol of sulfamic acid in $250\,cm^3$ volumetric flask = $M_r \times 0.025$, $97.1 \times 0.025 = 2.43\,g$ ✓ or similar. It should be made clear how the mass of solid used is determined (e.g. mass of weighing boat + solid before and after ✓ⓐ solution made up using distilled water ✓ⓐ.

ⓔ ⓐ Alternatives, but use of distilled water to make up solution can only be awarded once in this question.

(b) Discuss the suitability of each of the two standards used in this titration. You should comment on any errors in preparing each standard. (4 marks)

ⓔ In this context 'discuss' means examine the pros and cons of each solution as a standard. It may be that one standard *is* a better option than the other, but you need to make a comment about *both* possible standards. The second part of the question indicates that you should show an appreciation of any error inherent in the preparation of the standard.

(b) The HCl standard can only be as good as the original $1.00\,\mathrm{mol\,dm^{-3}}$ solution. The solution is less stable than the solid and thus changes in concentration: for example, evaporation of water from solution will change the concentration ✓. Solution is made using a $25\,\mathrm{cm^3}$ volumetric pipette, which has an accuracy of $\pm0.05\,\mathrm{cm^3}$ and thus percentage error = $(0.05/25) \times 100 = 0.2\%$ ✓.

ⓔ Any error calculation will score this mark.

Sulfamic acid is a stable solid and can be weighed to $\pm0.01\,\mathrm{g}$. Thus, assuming purity of the solid, we can be sure of the exact concentration of the standard ✓. However, percentage error in mass reading = $(0.01/2.43) \times 100 = 0.41\%$ ✓

Question 2*

This question concerns isomers of C_4H_9Br and their comparative rates of hydrolysis.

(a) Draw skeletal formulae for all bromoalkane isomers with the molecular formula C_4H_9Br. Name the isomers and classify each isomer as primary secondary or tertiary. (4 marks)

ⓔ This is a straightforward introduction to a practical-type question and should help you to focus on the particulars of the practical experiment.

Student answer

(a)

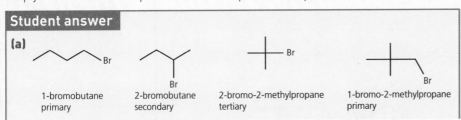

| 1-bromobutane | 2-bromobutane | 2-bromo-2-methylpropane | 1-bromo-2-methylpropane |
| primary | secondary | tertiary | primary |

ⓔ 1 mark for each correct structure (1 × 4); 1 mark each for each correct name *and* correct designation of primary/secondary/tertiary. Penalise displayed formula only once.

(b) Give details of an experiment that would enable you to compare the rates of hydrolysis of each bromoalkane with sodium hydroxide. You should include all necessary practical details, including any necessary measurements that must be made. (6 marks)

ⓔ You will probably have carried out this or a similar core practical (see CP4). It is certainly worth planning questions of this type very briefly in rough before committing your answer to the exam paper. Too many written answers are poorly planned and untidily presented.

(b) Equal volumes of each bromoalkane are measured out (this could also be equimolar quantities of each bromoalkanes using density and volume data) ✓. The other reagent is a fixed quantity of sodium hydroxide *in ethanol* (bromoalkanes are insoluble in just water) ✓. Silver nitrate added ✓. Time taken for precipitate to form is recorded ✓. Precipitate is silver bromide (AgBr) ✓. Temperature variable must be controlled ✓ for fair comparison (e.g. water bath).

Question 3*

Ethanol can be oxidised under carefully controlled experimental conditions to produce ethanal.

(a) Give details of how you could prepare and collect a sample of ethanal from a 20.0 cm³ sample of pure ethanol. You answer should include a labelled diagram of the apparatus and details of any reagents used. **(8 marks)**

ⓔ As with many questions of this type a neat, well-labelled diagram can score you many marks and *please* use a ruler!

If you have carried out core practical CP5, you will have oxidised ethanol to ethanoic acid. Although the reagents for converting ethanol to ethanal are the same, it is the careful modifications of the procedure that enable you to make ethanal the product.

Student answer

(a)

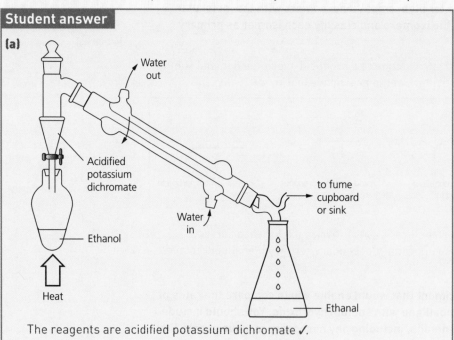

The reagents are acidified potassium dichromate ✓.

ⓔ Should be added from tap funnel to ethanol in flask.

By adding the acidified potassium dichromate dropwise from the tap funnel the oxidising agent is never in excess and thus as soon as ethanal is produced, it boils and is condensed ✓.

ⓔ Pear-shaped or round-bottomed flask ✓ correctly connected to condenser ✓. Apparatus must indicate pressure release 'breather tube' after condenser ✓. Water supply to condenser must indicate in at bottom out at top ✓. Anti-bumping granules in reaction flask ✓. Indication of heat source ✓.

(b) Give details of a *qualitative* experimental test that you could use to show that your product sample contains molecules with the aldehyde functional group. (4 marks)

(b) The sample can be tested with Tollens' reagent ✓. Silver nitrate and sodium hydroxide are added together to form a precipitate ✓. This precipitate is dissolved in ammonia solution ✓. The resultant solution is added to aldehyde and warmed. A positive test for the aldehyde functional group gives a silver mirror on the inside surface of the test tube ✓.

ⓔ As this question asks for details, simply saying Tollens' reagent gives a silver mirror will only score 2 marks out of the 3 available.

An alternative test is to use Fehling's solution(s) ✓. An alkaline solution of copper(II) ions ✓ added to an aldehyde gives an orange precipitate ✓ of copper(I) oxide ✓.

(c) Pure ethanol has a density of $0.789 \, g \, cm^{-3}$ and pure ethanal a density of $0.778 \, g \, cm^{-3}$. If $14.5 \, cm^3$ of ethanol was collected at the end of the experiment, calculate the percentage yield, stating any assumptions made. (4 marks)

ⓔ Your answer should include some indication of converting volume to mass.

(c) Using volume × density ✓, $20 \times 0.789 = 15.78 \, g$.

Mass to moles conversion. M_r of ethanol = 46, so $15.78/46 = 0.34(3)$ moles ✓. The same number of moles of ethanal can, in theory, be produced, so given M_r ethanol = 44, then $0.343 \times 44 = 15.09 \, g$, which has a volume of $15.09/0.778 = 19.40 \, cm^3$ ✓.

Thus percentage yield = (actual yield/theoretical yield) × 100 = (14.5/19.4) × 100 = 74.7% ✓.

ⓔ Accept 75%.

The assumption made is that the only product collected is ethanol ✓. Impurities such as water are possible.

Question 4*

Measuring the enthalpy of hydration using Hess's law.

A student wanted to measure the enthalpy change for the following reaction:

$$MgSO_4(s) + 7H_2O(l) \rightarrow MgSO_4.7H_2O(s)$$

(a) Explain why the enthalpy change for this reaction cannot be measured directly by adding pure water to anhydrous $MgSO_4(s)$ in the molar ratio 7:1. (2 marks)

ⓔ Measuring the enthalpy for many reactions is practically impossible, especially if the reaction is endothermic. Imagine putting a cake mixture in the oven and trying to find out *only the amount of energy taken in* as the cake bakes! The solution to the problem is to use Hess's law, which states that 'the enthalpy of reaction is independent of the route taken for the reaction'.

Student answer

(a) Water in the hydrated salt is in the form of water of crystallisation ✓, which means that there is an ordered arrangement of the water molecules in the crystal structure ✓. This cannot be reproduced by adding the measured amount of water to the anhydrous salt as it will simply dissolve a small amount of the anhydrous salt ✓.

ⓔ Any two.

(b) Give details of a method that you could use to measure the enthalpy change for this reaction indirectly. Include details of apparatus and any measurements made.

(8 marks)

ⓔ Again, a *labelled* diagram is useful in helping you to think about the procedure. Your diagram can gain credit for apparatus used.

(b)

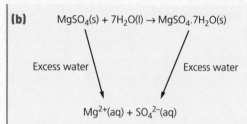

$$MgSO_4(s) + 7H_2O(l) \rightarrow MgSO_4.7H_2O(s)$$

Excess water   Excess water

$$Mg^{2+}(aq) + SO_4^{2-}(aq)$$

The Hess's law diagram for this reaction is shown above.

A calorimeter made of a suitable insulating material (e.g. polystyrene cup) ✓. A thermometer calibrated to measure to *at least* 1°C intervals ✓. A fixed mass/volume of water ✓ (e.g. 50–100 cm³). Mass of each solid should be recorded to 0.01 g ✓. Initial temperature of water taken ✓ (temperatures taken at 30 s intervals over 2 minutes to ensure a stable initial temperature).

Solid added to water *and* stirred with thermometer ✓ at time *t*. Temperatures recorded at fixed intervals (e.g. 30 seconds) ✓. The temperature should be recorded well after the maximum temperature is achieved to ensure rate of cooling can be measured ✓. An extrapolation of the rate of cooling can be made to give estimation of temperature change the instant the solid is added to the water ✓.

ⓔ Note that this last mark is almost certainly easier to attain if you *sketch* a graph of a typical dataset and show the extrapolation. Any 8 marks awarded from a possible 9 available.

(c) As a result of calculations made, the value for the enthalpy change is given as $\Delta H = -85.64\,kJ\,mol^{-1}$. Given that the thermometer is accurate to $\pm1°C$, present the answer to the suitable number of significant figures, explaining your reasoning. (2 marks)

> **(c)** Suitable value = $-86\,kJ\,mol^{-1}$ ✓. The final accuracy of your measurement can only be given to the *least* precise measurement ✓.

ⓔ As temperatures can only be recorded to 2 s.f., then final value can only be given to 2 s.f.

Question 5

An old sample of $FeSO_4.7H_2O$ is analysed to test its purity using the following method:

1 8.14 g of the solid is transferred into a beaker and dissolved in distilled water.

2 The solution is made up to 250 cm³ in a volumetric flask.

3 25 cm³ aliquots of this solution were titrated against 0.02 mol dm⁻³ acidified $KMnO_4(aq)$.

4 The mean titre was found to be 26.35 cm³ of $KMnO_4(aq)$.

(a) Use the two half-equations below to generate a full ionic equation for this reaction: (1 mark)

$$8H^+(aq) + MnO_4^-(aq) + 5e^- \rightarrow Mn^{2+}(aq) + 4H_2O(l)$$
$$Fe^{2+}(aq) \rightarrow Fe^{3+}(aq) + e^-$$

Student answer

(a) $8H^+(aq) + MnO_4^-(aq) + 5\,Fe^{2+}(aq) \rightarrow Mn^{2+}(aq) + 5Fe^{3+}(aq) + 4H_2O(l)$ ✓

(b) Describe *and* explain how the end point is determined in this reaction. (2 marks)

ⓔ It is important to pay attention to the 'command' words in a question. In this case you are asked to both describe and explain an observation. Put simply, this means first 'what happens?' and second 'why does it happen?'.

> **(b)** The end point is a pale pink colour ✓ due to the extra drop of $MnO_4^-(aq)$ *that is not reduced* ✓.

(c) Use this information to calculate the percentage purity of the $FeSO_4.7H_2O$ sample. (4 marks)

> **(c)** Moles of $KMnO_4$ in average titre = $(26.35/1000) \times 0.02 = 5.27 \times 10^{-4}$ ✓
>
> The balanced equation shows that for every 1 mole of potassium manganate(vii) reduced, 5 moles of $Fe^{2+}(aq)$ are oxidised.
>
> Thus $5 \times (5.27 \times 10^{-4}) = 2.635 \times 10^{-3}$ moles of Fe^{2+} oxidised ✓.
>
> This means 2.635×10^{-2} moles in 250 cm³ of solution ✓.

ⓔ Note here that we have taken the number of moles in the 25 cm³ titration and multiplied by 10 to get the total number of moles of Fe^{2+}(aq) in the 250 cm³ volumetric flask.)

$2.635 \times 10^{-2} \times 277.9$ (M_r of $FeSO_4.7H_2O$) = 7.32 g ✓

Thus percentage purity = 7.32/8.14 × 100 = 89.9% ✓.

(d) **Suggest why hydrochloric acid should not be used to acidify the potassium manganate(vii) solution. You may refer to your data booklet.** (2 marks)

ⓔ You may not always be asked to refer directly to the data booklet for a question but if you *can* support you answer with data, as in this case, then you demonstrate a clear understanding of the question.

(d) Acidified $KMnO_4$(aq) is a sufficiently strong oxidising agent ✓ (+1.51 V in data booklet) and would oxidise chloride ions to chlorine (data booklet +1.36 V) ✓.

ⓔ Consequently you would overestimate the amount of Fe^{2+}(aq) in the sample.

(e) **The table below shows the error inherent in each piece of apparatus in this experiment. Complete the table and calculate the total possible error in this method.** (3 marks)

Apparatus	Balance	Volumetric flask	Pipette	Burette
Precision error	±0.01 g	±0.06	±0.05	±0.05 (per reading)
Percentage error	....	....	....	....

ⓔ Calculate the error of each piece of apparatus and then add the errors together.

(e)

Apparatus	Balance	Volumetric flask	Pipette	Burette
Precision error	±0.01 g	±0.06	±0.05	±0.05 (per reading)
Percentage error	(0.01/8.14) × 100 = 1.2%	(0.06/250) × 100 = 0.02%	(0.05/25) × 100 = 0.2%	((0.05 × 2)/26.35) × 100 = 0.38%

Total error = 1.2 + 0.02 + 0.2 + 0.38 = 1.8% ✓

ⓔ Half a mark for each correct calculation in the table. Round up marks appropriately, e.g. 2.5 to 3 marks.

(f) **Use your answers to parts (c) and (e) to calculate the range for your percentage purity.** (1 mark)

(f) Range = 89.9 ± 1.8 ✓ = 88.1–91.7% ✓

Question 6*

A mistake has led to the mixing of samples of potassium carbonate and potassium sulfate in the same container. The container has approximately 50 g of solid.

Giving practical details of how you would carry out each part of the experiment, and *using full or ionic equations where appropriate*, design an experiment to:

(a) Identify the potassium cation, and the carbonate and sulfate anions in the mixture. (6 marks)

ⓔ This question requires you to distinguish between simple *qualitative* tests — what is there? — and *quantitative* tests — how much is there? If a question identifies a statement **in bold**, then you are required to address this part of the question if you are to gain full credit.

Student answer

(a) A qualitative test for potassium cation would be to carry out a simple flame test. Take a nichrome/platinum loop and clean it by putting if into a roaring flame. Ensure it is impurity free by dipping it into concentrated HCl and then flaming again ✓.

Take a small sample of the solid on the wire loop and place in a non-luminous flame. A lilac flame colour indicates the presence of potassium ions ✓.

Qualitative test for cations: carbonate. First take a large spatula of the solid (or any reasonable measured mass) and place in test tube. Add dilute HCl to the sample and bubble the gas released through lime water ✓. Lime water turning cloudy indicates the presence of the carbonate ion ✓.

$$2H^+(aq) + CO_3^{2-}(aq) \rightarrow H_2O + CO_2(g) ✓$$

ⓔ This mark can be awarded later in the qualitative analysis.

Test for sulfate: take a spatula of the solid sample and dissolve in distilled water ✓. Now add excess HCl(aq) until there is no further effervescence (it is important to remove all carbonate ions before testing for sulfate as barium carbonate will also give a precipitate) ✓. Add barium chloride solution ✓. A positive test is a white precipitate ✓.

$$Ba^{2+}(aq) + SO_4^{2-}(aq) \rightarrow BaSO_4(s) ✓$$

ⓔ You must include state symbols. This mark can be awarded later in the qualitative analysis. Any 8 from 10 possible marks.

(b) Quantify the percentage of each of the two potassium compounds in the mixture. You can assume there are no other impurities. (8 marks)

ⓔ It is better to calculate both quantities separately rather than just do one quantitative analysis and infer that the remaining mass is the other solid.

(b) Quantitative analysis: a measured mass of the solid must be recorded (suggest 2 g to 10 g maximum) ✓.

Collect the volume of gas (CO_2) produced when *excess* HCl is added to a measured quantity of the solid (the excess ensures that all of the carbonate reacts) ✓. The number of moles of CO_2 collected either over water or in a gas syringe ✓ is equal to number of moles of K_2CO_3 in the sample (see equation above) ✓.

ⓔ Marks for collection of CO_2 can be awarded from diagram.

Thus the proportion of K_2CO_3 in the sample can be calculated using number of moles = volume of gas in $cm^3/24\,000$ at standard temperature and pressure ✓.

ⓔ Note that for greater accuracy the volume must be adjusted using $PV = nRT$ for non-standard conditions.

To quantify the amount of potassium sulfate, a known mass of solid must be measured out. Excess HCl is then added to remove all carbonate ions ✓. Then excess barium chloride is added, so the precipitate forms.

ⓔ Again, the excess here ensures that all of the sulfate ions are precipitated out of solution.

The precipitate is removed by filtration ✓, washed and dried. The mass of precipitate is then found. Number of moles of $BaSO_4$ = number of moles of K_2SO_4 in the sample ✓. This can then be calculated as a percentage of initial mass ✓.

Question 7

Propan-1-ol reacts with ethanoic acid to produce an ester and water.

(a) Write a balanced equation for this reaction and name the organic product. (2 marks)

Student answer

(a) $CH_3COOH(l) + CH_3CH_2CH_2OH(l) \rightleftharpoons CH_3COOCH_2CH_2CH_3(l) + H_2O(l)$ ✓

The organic product is propylethanoate ✓.

(b) Write an expression for K_c for this reaction. (2 marks)

(b) $K_c = \dfrac{[CH_3COOCH_2CH_2CH_3(l)][H_2O(l)]}{[CH_3COOH(l)][CH_3CH_2CH_2OH(l)]}$

ⓔ Correct equation K_c ✓. Square brackets used ✓.

(c) You are required to find a value for K_c for this reaction at 298 K. Give details of a procedure that you would use to determine this value. You may use the data provided in the table. (7 marks)

	Ethanoic acid	Propan-1-ol
Density/$g\,cm^{-3}$	1.049	0.804

(c) Suitable volumes of acid and alcohol chosen (e.g. 20–25 cm^3) of each ✓. Each volume should be converted into number of moles using density data in table above ✓. A known number of moles of H_2SO_4 should be added to act as a catalyst ✓. The reaction is left for at least 5 days to reach equilibrium ✓. Once at equilibrium the amount of acid at equilibrium can be found by titrating against standardised NaOH ✓, using a suitable *named* indicator to measure the end point (e.g. phenolphthalein). The number of moles of acid added as the catalyst must be subtracted from the final number of moles of acid ✓. The quantities of the other reactants and products can be found from the stoichiometric equation. These values are inserted into the expression for K_c ✓.

ⓔ Because there are equal numbers of moles of reactants and products, the total volume does not change and thus molar quantities can be used in place of concentration values.

Question 8

In an experiment to prepare hydrated crystals of the double salt $Fe(NH_4)_2(SO_4)_2.6H_2O$ the following procedure was followed:

1 **Measure out 30 cm^3 of distilled water into a conical flask.**

2 **Measure out 4 cm^3 of concentrated sulfuric and add it to the water.**

3 **Slowly add 5.0 g of iron filings to the conical flask and stir while heating gently.**

4 **Add 10.0 g of ammonium sulfate, $(NH_4)_2SO_4(s)$, to the conical flask and allow it to dissolve.**

5 **Evaporate between half and two thirds of the volume of the solution and allow it to crystallise.**

(a) Suggest why the sulfuric acid is added to the water and not the water to the acid. (2 marks)

Student answer

(a) The reaction between the acid and the water is highly exothermic ✓. Water added to the acid would therefore boil quickly and splash dangerously ✓.

(b) Write an ionic equation, including state symbols, for the reaction between the acid and the iron filings. (2 marks)

(b) $Fe(s) + 2H^+(aq) \rightarrow Fe^{2+}(aq) + H_2(g)$

ⓔ Correct equation ✓, correct state symbols ✓.

(c) Using appropriate calculations, identify the limiting reagent in this procedure and calculate the maximum theoretical yield of the double salt. (4 marks)

(c) M_r of Fe = 55.8. Thus moles of Fe = 5.0/55.8 = 8.96×10^{-2} moles ✓

Number of moles of ammonium sulfate $(NH_4)_2SO_4$, M_r = 132.1 ✓, = 10/132.1 = 7.57×10^{-2} moles ✓

Formula of crystals is $Fe(NH_4)_2(SO_4)_2.6H_2O$, so iron is in excess and ammonium sulfate is the limiting reagent ✓.

So 7.57×10^{-2} moles formed. M_r salt = 356 g mol^{-1} ✓, so theoretical mass = 26.9 g ✓.

(d) If 24 g of dry salt crystals were collected at the end of the procedure, then calculate the percentage yield in this experiment. (1 mark)

(d) (24/26.9) × 100 = 89.2% yield ✓

Question 9*

Compound X is an organic liquid at room temperature and is composed of the elements C, H and O only. Combustion analysis of 4.50 g of compound X yields 6.60 g of $CO_2(g)$ and 2.70 g of $H_2O(l)$.

(a) Calculate the empirical formula of X. (4 marks)

ⓔ The masses of carbon and hydrogen can be found from the masses of carbon dioxide and water produced in the combustion reaction. The total mass of these two elements minus the original mass will give you the mass of the oxygen in X.

Student answer

(a) 6.60 g of CO_2 = 6.6/44 = 0.15 mole of CO_2, so 0.15 mole of C in 4.50 g of sample ✓.

2.70 g of H_2O = 2.7/18 = 0.15 moles of H_2O ✓, so 0.3 moles of H in sample ✓.

Total mass of C and H = (0.15 × 12) + (0.3 × 1) = 2.1 g. Thus mass of oxygen = 4.5 – 2.1 = 2.4 g ✓.

ⓔ Now that we have the masses of each element we can convert to a molar ratio and thus an empirical formula.

Moles of O = 2.4/16 = 0.15 ✓. Thus the formula is $C_{0.15}H_{0.3}O_{0.15}$ or CH_2O ✓.

(b) The molecular mass of X is twice the empirical formula. Calculate the molecular formula of X. (1 mark)

(b) Molecular formula = $C_2H_4O_2$ ✓

ⓔ Note that '$2CH_2O$' is *incorrect* as this means two moles of a compound with the formula CH_2O.

(c) Draw and name two **functional group isomers** of compound X. (4 marks)

(c)

Ethanoic acid   Methyl methanoate   Ethene-1,1-diol
(also *E* and *Z* isomers of ethene-1,2-diol)

ⓔ Any two correct structures (✓); can be displayed formula. Any two correct names (✓). Other structures are possible, such as isomers of ethenediol.

(d) An isomer of X reacts with sodium carbonate to produce a salt, water and carbon dioxide. Write a balanced equation for this reaction and name the isomer. (2 marks)

(d) $2CH_3COOH + Na_2CO_3 \rightarrow 2CH_3COO^-Na^+ + H_2O + CO_2$ ✓

The isomer is ethanoic acid ✓.

ⓔ Equation with correct reactants ✓, correct products ✓ (state symbols not required). Do not give information that is *not* asked for in the question.

Question 10

Sodium thiosulfate reacts with hydrochloric acid according to the following equation:
$$Na_2S_2O_3(aq) + 2HCl(aq) \rightarrow 2NaCl(aq) + SO_2(g) + S(s) + H_2O(l)$$

You are provided with solutions of $1.0\,mol\,dm^{-3}$ HCl(aq) and $0.05\,mol\,dm^{-3}$ $Na_2S_2O_3(aq)$ and standard laboratory apparatus.

(a) Design an experiment to investigate the effect of changing temperature on this reaction. Your method should include:
- suggested quantities of reagents and suitable apparatus
- independent, dependent and controlled variables
- any safety precautions you should take
- all measurements taken (6 marks)

ⓔ Always pay attention to the state symbols in a given equation.

Student answer

(a) In this experiment a precipitate of sulfur is formed as the reaction progresses. It is not practically possible to determine when the reaction completely stops, but it is possible to determine the time when the reaction has reached the same point by using the time taken for a cross drawn on a piece of paper and placed under the reaction vessel (e.g. a conical flask) to be completely obscured by the sulfur produced.

Suitable apparatus and quantities would be a $100\,cm^3$ conical flask ✓ and adding $20\,cm^3$ volumes of each reagent together ✓.

Questions & Answers

e You should try suitable quantities for your chosen piece of apparatus. It is always worth trying out quantities to see what works practically.

> The temperature of both reagents should be taken prior to adding them together and the time taken for the cross to 'disappear' recorded ✓. Various temperatures can be investigated using a temperature-controlled water bath and/or ice to achieve a chosen range of temperatures ✓. The volumes and concentrations of both reagents should be kept constant throughout the investigation ✓. Note that in this experiment sulfur dioxide gas is produced and thus for safety reasons the reaction should be carried out in a fume cupboard or in a well-ventilated area ✓.

(b) Complete the table. (3 marks)

Experiment	Time/s	$\log_e t$	Temperature/K	$\times 10^{-3} K^{-1}$	$\log_e K^{-1}$
1	102		290		
2	61		298		
3	36		306		
4	25		313		
5	17		318		
6	11		327		

e 1 mark is awarded per fully correct column. Make sure that you can complete the remaining columns of the table correctly using the correct functions on your calculator. It is especially important that you use the correct logarithmic function usually referred to as 'ln' on your calculator.

(b)

Experiment	Time/s	$\log_e t$	Temperature/K	$\times 10^{-3} K^{-1}$	$\log_e K^{-1}$
1	102	4.62	290	3.45	−5.67
2	61	4.11	298	3.36	−5.70
3	36	3.58	306	3.27	−5.72
4	25	3.22	313	3.19	−5.75
5	17	2.83	318	3.14	−5.76
6	11	2.40	327	3.06	−5.79

(c) A value for the activation energy for this reaction can be calculated from the following relationship:

time $(t) \propto Ae^{+E_a/RT}$

where E_a = activation energy. Use the data and a suitable plot to calculate a value for E_a in this reaction. (4 marks)

(c)

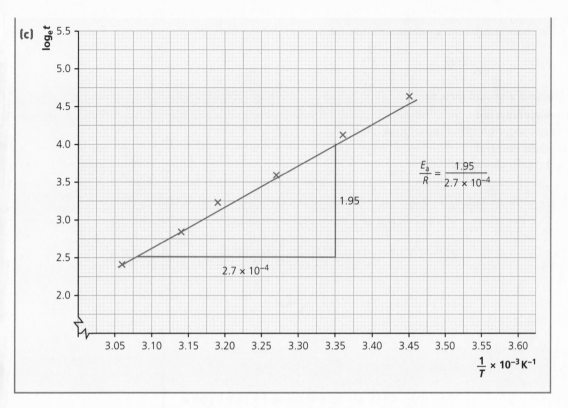

$$\frac{E_a}{R} = \frac{1.95}{2.7 \times 10^{-4}}$$

ⓔ Correct x-axis is $1/T \times 10^{-3}\,K^{-1}$ ✓, correct y-axis is $\log_e t$ ✓, all points correctly plotted (✓) (subtract 1 for each incorrect plot, max −2). Correct line of best fit ✓.

Gradient of line = E_a/R = $1.95/(2.7 \times 10^{-4})$. Taking R as $8.31\,J\,K^{-1}\,mol^{-1}$ then E_a = $60\,kJ\,mol^{-1}$✓. You will be given a mark based on the gradient of the line you have drawn.

Question 11*

Cyclohexene can be prepared by the catalytic dehydration of cyclohexanol, separated and subsequently dried using a suitable drying agent.

(a) (i) Write a balanced equation for the reaction. (1 mark)

Student answer

(a) (i) $C_6H_{11}OH(l) \rightarrow C_6H_{10}(l) + H_2O(l)$ ✓

(ii) Calculate the atom economy for the synthesis. (1 mark)

(ii) Total mass of reactant = 100, M_r of C_6H_{10} = 82. Thus atom economy = 82% ✓.

(b) You are provided with a $30\,cm^3$ sample of cyclohexanol, standard laboratory glass apparatus and standard laboratory reagents. Outline a procedure for the synthesis and collection of a pure dry sample of cyclohexene. You should refer to the data in the following table in your answer.

	Cyclohexanol	Cyclohexene
Boiling point/°C	161	84
Density/g cm^{-3}	0.962	0.811

Your procedure should include:

- the reagents used
- a clear indication of the techniques used
- a suitable catalyst
- a suitable drying agent

(8 marks)

ⓔ In answering this type of question it is important that you address each of the bullet points specified in the question. After drafting an answer in rough, check it over and tick off the bullet points as you address them in your answer.

(b) The reactant is cyclohexanol and the acid catalyst is (concentrated) phosphoric(v) acid ✓.

ⓔ Concentrated sulfuric acid is *not* a good option for this reaction as it is a powerful oxidising agent.

Cyclohexanol and phosphoric acid are added to a reaction flask (pear-shaped or round-bottomed on ice) as the reaction is exothermic ✓. Anti-bumping granules are added to the flask ✓. Mixture heated under reflux ✓ (for between 15 and 30 minutes). The apparatus is dismantled and set up for distillation ✓. Distil off cyclohexene from the reaction mixture (keeping temperature below 100°C). Add distillate to the separating funnel and then add sodium chloride solid ✓ until no more will dissolve. Discard the lower aqueous layer and then redistil sample between 81°C and 86°C (refer to the table) ✓. Collect final sample in a stoppered flask and add a small spatula of a named suitable drying agent (solid calcium chloride, anhydrous sodium sulfate) ✓.

(c) Infrared spectroscopy can be used to assess the purity of your product. What might a small broad peak above 3100 cm^{-1} (refer to data booklet) suggest about your product? Explain your answer. (2 marks)

(c) The peak indicates an O–H bond ✓. This suggests either some unconverted cyclohexanol or traces of water present in the sample ✓.

Question 12

The rate equation for the reaction between acidified hydrogen peroxide and iodide ions can be investigated using an 'iodine clock' reaction.

The equation for the reaction is:

$$H_2O_2(aq) + 2I^-(aq) + 2H^+(aq) \rightarrow I_2(aq) + 2H_2O(l)$$

A time t from which a rate measurement can be made is determined by the following reaction:

$$2S_2O_3^{2-}(aq) + I_2(aq) \rightarrow S_4O_6^{2-}(aq) + 2I^-(aq)$$

A starch indicator is used to determine the end point for this reaction.

(a) What observation is made at the end point for this reaction? (1 mark)

Student answer

(a) Appearance of a blue/black colour ✓ due to iodine–starch precipitate.

(b) Why does the stoichiometric equation for the reaction suggest that the reaction *must* involve a number of steps? (2 marks)

(b) The stoichiometric equation shows five reactant molecules ✓. The probability of an instantaneous collision between this exact number of molecules in these proportions with sufficient activation energy is vanishingly small, so there must be more than one step ✓.

(c) In a trial run a student uses the quantities of solutions shown in the table.

Volume of $0.25\,mol\,dm^{-3}$ sulfuric acid/cm^3	Volume of $0.1\,mol\,dm^{-3}$ hydrogen peroxide/cm^3	Volume of $0.1\,mol\,dm^{-3}$ potassium iodide/cm^3	Volume of $0.05\,mol\,dm^{-3}$ sodium thiosulfate/cm^3	Volume of starch indicator/cm^3
25	10	10	25	1

With reference to the equations shown explain, using appropriate calculations, why the student would not have observed an end point for these quantities. (3 marks)

ⓔ This question requires you to have a clear understanding of the two reactions taking place. Iodide oxidised by the hydrogen peroxide to iodine is immediately converted back to iodide *as long as there is thiosulfate present*. When the thiosulfate is used up, the iodine remains to form the blue/back precipitate.

(c) In the student's trial run, there are $(10/1000) \times 0.1 = 1 \times 10^{-3}$ moles of iodide ✓, which can form 5×10^{-4} moles of iodine ✓. There are $(25/1000) \times 0.05 = 1.25 \times 10^{-3}$ moles of thiosulfate ✓. Thus the thiosulfate is in excess ✓ and thus there will be no end point.

Question 13

(a) Outline a method for measuring the standard potential difference generated when the two half-cells shown below are connected:

$$Zn^{2+}(aq) + 2e^- \rightleftharpoons Zn(s) \qquad -0.76\,V$$
$$Fe^{3+}(aq) + e^- \rightleftharpoons Fe^{2+}(aq) \qquad +0.77\,V$$

Your method should include the following:
- a labelled diagram
- details of how the two half-cells can be prepared (specific quantities are not required)
- how the salt bridge is prepared (9 marks)

Student answer

(a)

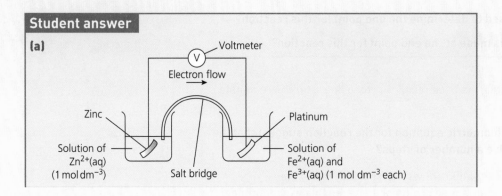

e Once again, a clearly drawn and well-labelled diagram will score you most of the marks for this part of the question.

A (high-resistance) voltmeter ✓ connected to Zn(s) electrode ✓ in solution of Zn^{2+}(aq) ions of concentration $1.0\,mol\,dm^{-3}$ ✓. Other half-cell, platinum electrode ✓ in solution that is $1.0\,mol\,dm^{-3}$ for both Fe^{2+}(aq) ✓ and Fe^{3+}(aq) ✓. Salt bridge ✓ made from filter paper (or other porous material) ✓ soaked in ionic solution such as potassium nitrate ✓.

(b) Write an equation for the overall reaction and calculate the expected E_{cell} for this reaction. (2 marks)

(b) $Zn(s) + 2Fe^{3+}(aq) \rightarrow Zn^{2+}(aq) + 2Fe^{2+}(aq)$

e Correct reactants and products, and correct balancing of equation. ✓

$E_{cell} = 0.77 - (-0.76) = 1.53\,V$ ✓

(c) Indicate on your diagram the direction of electron flow in the circuit. (1 mark)

e Electron flow is from Zn half-cell to Fe^{2+} half-cell ✓ (shown on diagram).

Question 14

Compound Q

(a) (i) Give the molecular formula for Q. (1 mark)

Student answer

(a) (i) C_5H_8O ✓

(ii) Give the full systematic name for compound Q.   (2 marks)

🅔 If a molecule contains a C=C bond, then be aware that there may be geometric isomers and thus you should be ready to use the *E/Z* isomerism nomenclature, as in this case.

(ii) *Z*-prop-3-enal ✓✓

🅔 Name ✓; *Z* isomer ✓.

(b) Give details of the tests that you could use to identify the functional groups present in Q and the expected results of each test.   (5 marks)

(b) Test for alkene group ✓. Bromine water *in the absence of direct sunlight* ✓ goes from orange to colourless ✓. Aldehyde group ✓. Ammoniacal silver nitrate solution (accept Tollens' reagent here) ✓ gives a silver mirror ✓ *or* Fehling's solution ✓ gives an orange precipitate ✓.

(c) Different oxidising agents will react with different functional groups in compound Q. Draw the skeletal formula and name the organic product produced when Q reacts with:

(i) acidified potassium dichromate(VI)   (2 marks)

(c) (i)

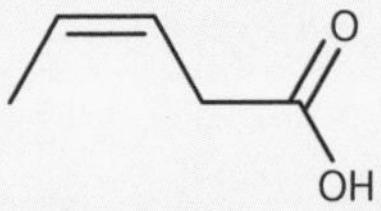

Z-prop-3-enoic acid

🅔 Correct structure ✓; correct name ✓.

(ii) alkaline potassium manganate(VII)   (2 marks)

(ii)

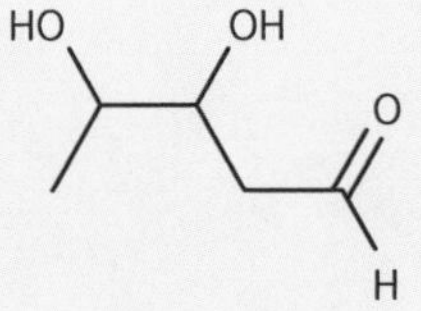

3,4-dihydroxypentanal

🅔 Correct structure ✓; correct name ✓.

Question 15

You are provided with a $0.1\,mol\,dm^{-3}$ solution of benzoic acid (a weak organic acid) and a standardised solution of $0.100\,M$ NaOH.

The K_a value of benzoic acid $= 6.3 \times 10^{-5}\,mol\,dm^{-3}$.

(a) Give an expression for K_a for benzoic acid.   (2 marks)

Student answer

(a) $K_a = \dfrac{[C_6H_5COO^-(aq)][H^+(aq)]}{[C_6H_5COOH(aq)]}$

ⓔ Numerator correct ✓; denominator correct ✓.

(b) Using the two solutions provided and standard titrimetric apparatus, outline a procedure that would enable you to prepare a buffer solution of pH = 4.2. You should illustrate your answer with relevant chemical equations. Remember that if $K_a = [H^+(aq)]$, then $pK_a = pH$. (5 marks)

ⓔ The key to this question is that pH = 4.2 is the same as $[H^+(aq)] = 6.3 \times 10^{-5}\,\text{mol dm}^{-3}$. This means that at the point of *half neutralisation* we will have a buffer of pH = 4.2. This is because we will have neutralised exactly half of the benzoic acid, giving equal concentrations of benzoic acid molecules and benzoate ions, as shown in the equation given in the answer below.

(b) $[H^+(aq)] = K_a \times \dfrac{[C_6H_5COOH(aq)]}{[C_6H_5COO^-(aq)]}$ ✓

Because the concentration of both benzoic acid and benzoate are the same, $[C_6H_5COOH(aq)]/[C_6H_5COO^-(aq)] = 1$, and so $[H^+(aq)] = K_a$ or $pK_a = pH$ ✓.

Pipette 25 cm³ of the benzoic acid solution into a conical flask ✓ and add a few drops of phenolphthalein indicator ✓. Titrate against the standardised NaOH solution until a faint pink end point is seen ✓.

ⓔ This will enable you to calculate the exact concentration of the benzoic acid solution.

Now measure out a volume of the benzoic acid and, using the results from the previous titration experiment, add exactly sufficient moles of NaOH to neutralise exactly half of the benzoic acid ✓. Test the pH of the resultant buffer with a calibrated pH meter.

Question 16*

Copper forms two oxides: Cu_2O and CuO.

Copper(II) carbonate decomposes on heating to form one of these oxides. Equations representing these possible decompositions are shown below:

$$4CuCO_3(s) \rightarrow 2Cu_2O(s) + 4CO_2(g) + O_2(g)$$
$$CuCO_3(s) \rightarrow CuO(s) + CO_2(g)$$

(a) Assign oxidation numbers to copper in each of the two oxides. (1 mark)

Student answer

(a) Cu_2O, $Cu = +1$; CuO, $Cu = +2$ ✓

ⓔ Both must be correct for the mark.

(b) Design an experiment to investigate which of these two decompositions takes place. Your method should include the following:

- a labelled diagram of the apparatus you will use
- suggested quantities of copper(ɪɪ) carbonate to be used
- measurements to be taken
- some sample calculations to indicate how you would process your data
- any precautions taken to ensure safety and a fair test

1 mole of any gas occupies 24 000 cm³ at 298 K and 101 000 Pa.   (12 marks)

ⓔ Again, this is a procedure that requires careful planning and thus a draft should be attempted before committing your final answer to the exam paper.

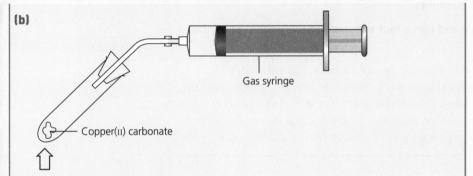

(b)

Gas syringe

Copper(ɪɪ) carbonate

ⓔ Marks can be awarded from the diagram: test tube/boiling tube containing sample ✓ connected to gas syringe ✓. Note here that a gas syringe is better than collecting over water as carbon dioxide is quite soluble in water.

> The equations show that the molar ratios for carbonate to gas are 1:1.25 for the first equation and 1:1 for the second ✓.

ⓔ As the gas syringe has a volume of 100 cm³ (or different *if* specified) then calculation of suitable amount of $CuCO_3(s)$ should be shown. For example:

> 100 cm³ = 100/24 000 = 4.17 × 10⁻³ moles. So the maximum number of moles of $CuCO_3(s)$ should be less than this ✓. $CuCO_3$ M_r= 123.5 g mol⁻¹ ✓. (4.17 × 10⁻³) × 123.5 = 0.515 g ✓ (or similar calculation).
>
> Thus, allowing for a 1:1.25 ratio a mass of 0.3–0.4 g would be suitable ✓.
>
> Record room temperature ✓. Add specified mass to the test tube and connect to the gas syringe (ensuring no leaks) ✓. Heat the sample until there is no further expansion of gas in the syringe ✓. Allow the apparatus to cool to room temperature ✓. Measure the final volume of gas collected ✓.

ⓔ The correct equation can be determined from the molar solid:gas ratio observed; i.e. 1:1.25 or 1:1.

> Repeat experiment ✓.

ⓔ Acknowledgment of any correction needed using $pV = nRT$ ✓. Any 12 marks from a possible 14.

Question 17

To have the best chance of answering this question well, read question parts (a), (b) and (c) before you start.

You are provided with $25\,cm^3$ samples of four different organic molecules. Each compound contains:

- **four carbon atoms**
- **eight *or* ten hydrogen atoms**
- **one *or* two oxygen atoms**
- **no other atoms of other elements**

(a) Draw skeletal formulae and name four organic molecules that fit all the above criteria.   (8 marks)

ⓔ If you have read the whole question before starting, you will choose your answers to this part of the question in such a way that will help you answer the remaining parts of the question. It is clearly much easier to distinguish molecules with different functional groups, so you should consider this before drawing your isomers. Four possible isomers are shown in the answer, but others are possible.

Student answer

(a)

| 2-methylpropan-2-ol | Butanone | Butan-1-ol | Butanoic acid |

ⓔ Each correct structure ✓; each correct name ✓.

(b) Devise a practical procedure that would enable you to identify each compound you have drawn in part (a). You should devise a clear, logical sequence of tests, including reagents and expected results.   (6 marks)

ⓔ There are a number of possible ways of tackling this question but some are easier to perform than others. You should think about a test that will give a positive for only one of the liquids each time. Thus you will be able to eliminate each one by one. One suggested procedure is given.

(b) For each test use 2–$3\,cm^3$ of each liquid.

Test all with solid sodium carbonate ✓. Only butanoic acid from those above would effervesce ✓. Test all three remaining samples with acidified potassium dichromate and warm ✓. Only butan-1-ol will be oxidised, turning the dichromate from orange to green ✓. Test the remaining two samples with a small amount of sodium metal ✓. Only the 2-methylpropan-2-ol of the two remaining will effervesce ✓, producing hydrogen. The remaining liquid is butanone.

(c) Draw skeletal formulae and name two organic molecules that fit the criteria stated above that you would *not* be able to distinguish using standard laboratory tests, and explain your reasoning. (4 marks)

ⓔ If functional group isomers are reasonably easy to distinguish, stereoisomers and chain isomers are much more difficult. Two possible structures are shown below.

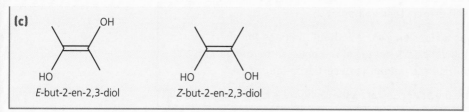

(c)

E-but-2-en-2,3-diol   *Z*-but-2-en-2,3-diol

ⓔ Correct structures fitting criteria in question stem ✓✓; correct names ✓✓.

Question 18

You are provided with a solution of approximately $0.05\,mol\,dm^{-3}$ ethanedioic acid $(COOH)_2$.

The exact concentration of this solution can be determined using two different types of titration:

- an acid–base titration using standardised NaOH(aq)
- a redox titration using acidified $KMnO_4$(aq)

(a) Write fully balanced stoichiometric equations for each of the two reactions. (4 marks)

ⓔ Remember when balancing full ionic equations that the charges must be the same on both sides.

> **Student answer**
>
> **(a)** $2NaOH(aq) + (COOH)_2(aq) \rightarrow (COO^-Na^+)_2(aq) + 2H_2O(l)$
>
> $5(COOH)_2(aq) + 2MnO_4^-(aq) + 6H^+(aq) \rightarrow 10CO_2(g) + 2Mn^{2+}(aq) + 8H_2O(l)$

ⓔ Correct reactants and products ✓; balanced ✓.

(b) Discuss the *two* methods for the determination of the concentration of the ethanedioic acid solution and suggest which method would give the more accurate result. (5 marks)

ⓔ Any titrimetric analysis is determined by the accuracy of the standard solution. In this question consider the two standardising solutions and think why one might be better than the other.

(b) The acid–base titration is dependent on the stability of the sodium hydroxide solution as a standard. NaOH is not a good standard as it will react with CO_2 in the atmosphere ✓:

$$2NaOH(aq) + CO_2(g) \rightarrow Na_2CO_3(aq) + H_2O(l)$$

However, the reaction goes to completion quickly and a suitable indicator, e.g. phenolphthalein, will give a clear end point ✓.

The redox titration gives a clear end point without the need for an indicator ✓. However, the very dark colour of the potassium manganate titrant makes it more difficult to judge the burette readings ✓.

Redox titration would give the better result as the standard is more stable in air ✓.

Question 19*

Five unlabelled bottles contain approximately $150 \, cm^3$ of the following substances:

- cyclohexane
- aqueous saturated calcium hydroxide solution
- nitric acid (approximately $2 \, mol \, dm^{-3}$)
- aqueous sodium iodide solution
- aqueous potassium chloride solution

(a) Devise a logical sequence of tests that would enable you distinguish the five substances. You have standard laboratory chemicals but are not allowed to use any pH indicators. (8 marks)

ⓔ Think about devising a sequence of tests that allows you to eliminate one substance at a time.

Student answer

(a) Test 1. Test all solutions by adding drops of each sample to see if any are immiscible with water ✓. Cyclohexane will form a separate phase on top of the water ✓. Could also test a small sample for combustibility in a fume cupboard. Only cyclohexane will combust.

Test 2. Prepare a stream of carbon dioxide gas and bubble through the remaining samples ✓. Saturated calcium hydroxide will give a cloudy precipitate ✓.

Test 3. Add a small quantity of powdered calcium carbonate to each solution ✓. Only nitric acid will effervesce ✓ and give off $CO_2(g)$.

Test 4. Add a few drops of nitric acid followed by a few drops of silver nitrate to the remaining samples ✓. Both will form precipitates ✓.

Test 5. By adding aqueous ammonia to each precipitate the silver chloride precipitate formed with the potassium chloride will dissolve ✓, but the silver iodide will remain insoluble ✓.

(b) Choose *two* of the substances you have identified and suggest suitable hazard symbol labels for the bottles. (2 marks)

(b) Cyclohexane — flammable ✓. Nitric acid — corrosive ✓.

Question 20*

You are provided with 250 cm^3 of approximately 1 mol dm^{-3} hydrochloric acid, 25 cm^3 of 1.00 mol dm^{-3} standardised sodium hydroxide solution and standard laboratory apparatus.

(a) Devise a procedure to determine the enthalpy of neutralisation for the reaction between hydrochloric acid and sodium hydroxide. Your procedure should include:
   - a labelled diagram of the apparatus used
   - an *ionic* equation representing the enthalpy of neutralisation
   - details of suggested quantities
   - measurements taken
   - details of how your data can be used to determine the enthalpy of neutralisation

   Assume all solutions have a standard heat capacity of 4.18 J mol^{-1} K^{-1}. (10 marks)

Student answer

(a)

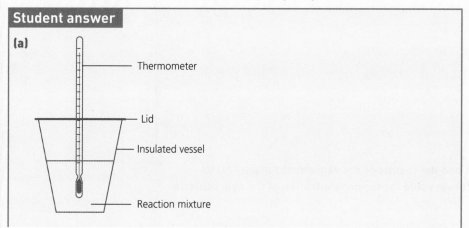

Thermometer

Lid

Insulated vessel

Reaction mixture

ⓔ Marks from diagram — suitable insulating vessel ✓, thermometer ✓, lid ✓.

$H^+(aq) + OH^-(aq) \rightarrow H_2O(l)$ ✓

Fill rinsed burette with HCl(aq) ✓. Pipette 25 cm^3 of standardised NaOH(aq) into reaction vessel ✓. Record temperature to nearest calibration possible every 30 seconds for 2 minutes ✓.

Add acid in 5 cm^3 volumes to the reaction mixture, stir and record temperature ✓. Add a further 5 cm^3 of acid and again record temperauture and repeat until 50 cm^3 of acid have been added ✓.

Plot points on graph and extrapolate both rising and falling temperature lines, to give a maximum temperature rise ✓.

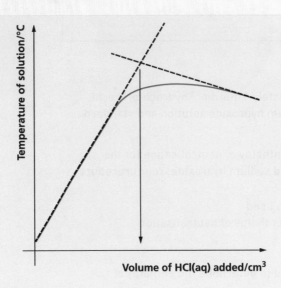

Processing of data:

Number of moles of water produced determined by number of moles of standard, i.e. 25 cm³ of 1.00 mol dm⁻³ solution = 0.025 moles ✓.

Temperature rise from experiment = ΔT.

Energy released = (total mass of solutions at maximum T (see graph)) × 4.18 × ΔT.

This value/0.025 = total energy released per mole of water formed. Sign will be negative as reaction is exothermic.

(b) **Explain how you could use the results of the experiment in part (a) to determine a more accurate value for the concentration of the hydrochloric acid solution.**

(3 marks)

(b) The point on the graph where the two extrapolated lines intersect is extended down to the x-axis ✓. The volume of HCl(aq) added at neutralisation can be read off ✓. This volume will contain exactly 0.025 moles of acid ✓, and the concentration of the acid can be found from concentration = 0.025/volume of acid added in dm³ ✓.

Question 21

In this experiment you will prepare a sample of 1-bromobutane from butan-1-ol.

$$CH_3CH_2CH_2CH_2OH + HBr \rightarrow CH_3CH_2CH_2CH_2Br + H_2O$$

The hydrogen bromide will be prepared *in situ* in the following reaction:

$$NaBr(s) + H_2SO_4 \rightarrow HBr(g) + NaHSO_4$$

(a) **Explain what is meant by the term *'in situ'* and suggest why this method might be preferred.**

(2 marks)

Student answer

(a) A reagent produced *in situ* is one where a separate reaction produces a reagent necessary for the main reaction in the same reaction vessel ✓. This can be a safer method especially when the reagent is hazardous or difficult to handle ✓.

(b) Calculate the *overall* atom economy for this reaction. (2 marks)

(b) $CH_3CH_2CH_2CH_2OH + HBr \rightarrow CH_3CH_2CH_2CH_2Br + H_2O$

$NaBr(s) + H_2SO_4 \rightarrow HBr(g) + NaHSO_4$

Overall: $CH_3CH_2CH_2CH_2OH + NaBr(s) + H_2SO_4 \rightarrow CH_3CH_2CH_2CH_2Br + H_2O + NaHSO_4$ ✓

Total mass of reactants = 191; mass of $CH_3CH_2CH_2CH_2Br$ = 137

So atom economy = 137/191 × 100 = 72% (71.7) ✓

You are provided with the data shown in the table.

	1-Bromobutane	Butan-1-ol	Concentrated HCl
Density/g cm^{-3}	1.3	0.81	1.2
Boiling point/°C	117	102	N/A
Solubility in water	Insoluble	Soluble	Soluble

Note that butan-1-ol is soluble in concentrated hydrochloric acid, but 1-bromobutane is not.

(c) Use your knowledge of organic synthesis with the data in the table to outline a method to prepare a pure, dry sample of 1-bromobutane from butan-1-ol. For the purposes of this experiment you may assume an 80% yield. Your method should include:

- A suggested quantity of butan-1-ol that would enable you to prepare about 5 g of 1-bromobutane. (You have 8 g of sodium bromide and 10 cm^3 of concentrated sulfuric acid, both of which are in excess.)
- Safety procedures that must be considered at each step.
- A labelled diagram of the apparatus that should be used at each step.
- How the product is separated from the final mixture.
- How the final product is dried and collected. (12 marks)

e A question of this type will require some considerable preparation. You need to be familiar with all of the key steps of synthesis and purification for a procedure such as this. Normally there will be more possible ways of securing marks than the total number of marks in the question. In the specimen answer below there are a total of 15 possible marking points, but a maximum of 12 for the question.

Making the hydrogen bromide *in situ* in a safe way requires that some thought be given to how the reactants are combined. By adding the concentrated sulfuric acid from a tap funnel, we can control the reaction by adding the acid dropwise to control the rate of this exothermic reaction. When necessary, the reaction can be warmed. The reflux condenser is set up to prevent any loss of reactants or products.

You should show preliminary calculations.

(c) 5 g of 1-bromobutane = 5/137 (M_r of 1-bromobutane) = 0.365 mol ✓. Synthesis gives only 80% yield, so based on 1:1 reaction ratio we need at least 0.365 × 100/80 mol = 0.0456 mol ✓.

Thus we will use 0.05 mol of butan-1-ol, which is 0.05 × 74 = 3.7 g. Butan-1-ol density = 0.81 g cm^{-3}, so volume = 3.7/0.81 = 4.57 cm^3. Use 5 cm^3 of butan-1-ol ✓.

ⓔ 5 cm^3 is a sensible quantity based on the information given and the laboratory apparatus available, but be sure to show this by calculation.

First dissolve the 8 g of sodium bromide in the minimum amount of distilled water. This is added to the round-bottom flask ✓ with some anti-bumping granules ✓. The measured volume of butan-1-ol can be added to the flask and the contents mixed ✓. Connect the apparatus as shown in Diagram 1.

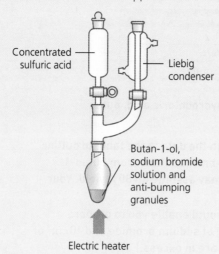

Concentrated sulfuric acid

Liebig condenser

Butan-1-ol, sodium bromide solution and anti-bumping granules

Electric heater

Diagram 1

The 10 cm^3 of concentrated acid is added dropwise from the tap funnel to control the rate of reaction ✓. Once all of the acid has been added, the mixture is heated under reflux for about 30 mins ✓ using an electric heater. Using the density of the 1-bromobutanol, mass/density, 5/1.3 = 3.8 cm^3. We should now reconnect the apparatus to distil over a volume of between 4 cm^3 and 5 cm^3 of product ✓. This is the crude product. See Diagram 2.

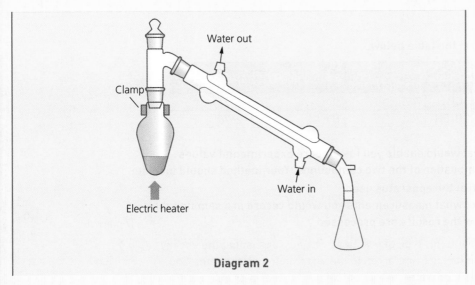

Diagram 2

e Remember to show how the water enters the condenser (from the bottom) and exits. There is frequently a mark for stating this.

The crude product is now placed in a separating funnel ✓ (see Diagram 3) and, using the information from the table, an equal volume of concentrated HCl ✓ is added to the separating funnel. The stop is placed on the funnel and the funnel carefully inverted a few times to mix the contents. Any unreacted butan-1-ol will dissolve in the HCl. The excess acid can be neutralised by adding small quantities of aqueous sodium hydrogen carbonate ✓, stopping and inverting the separating funnel, taking care to release the pressure due to any carbon dioxide produced ✓. This should be done until no more gas is produced. You can now tap off the lower phase ✓ (see table) and discard the upper phase. This sample is now put into a small conical flask with a spatula of a drying agent such as anhydrous calcium chloride ✓ and left overnight.

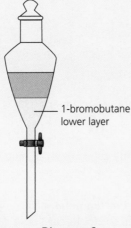

1-bromobutane
lower layer

Diagram 3

The final product can then be distilled off at a temperature between 101°C and 103°C ✓.

Question 22*

Look carefully at the data in the table below.

Compound	Structural formula	Molar mass/g mol^{-1}	Boiling point/°C	$\Delta_c H^{\ominus}$/kJ mol^{-1}
Ethoxyethane	$CH_3CH_2OCH_2CH_3$	74.1	35	−2724
Butan-1-ol	$CH_3CH_2CH_2CH_2OH$	74.1	117	−2676

(a) Describe a method that would enable you to compare experimental values for the enthalpy of combustion of the two compounds. Your method should include:

- a labelled diagram of the apparatus used
- a clear indication of what measurements you would record in a sample table
- an indication of how the results are processed   (7 marks)

ⓔ You should be familiar with this type of practical. The key idea is that the energy from the combustion of the fuel is transferred to the water in the calorimeter, and from the temperature rise of the water we can calculate the energy transferred. However, as no energy transfer is 100% efficient, you need to be aware that your experimental value is unlikely to be very close to the databook value.

Student answer

(a)

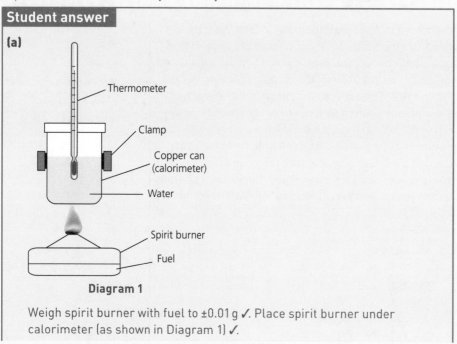

Diagram 1

Weigh spirit burner with fuel to ±0.01 g ✓. Place spirit burner under calorimeter (as shown in Diagram 1) ✓.

ⓔ Note that the distance between the spirit burner and the calorimeter should be kept constant.

Add 100 cm^3 of water to the calorimeter ✓. Place the thermometer inside the calorimeter and allow the temperature reading to settle. Record the initial temperature ✓. Light the burner and heat the water for a period of about 2 minutes. (In this period the water should not boil.) After 2 minutes, extinguish the burner and record the maximum temperature attained by the water ✓. Reweigh the spirit burner ✓.

	Ethoxyethane	Butan-1-ol
Initial mass of spirit burner/g		
Final mass of spirit burner/g		
Change in mass Δm/g		
Final temperature of water/°C		
Initial temperature of water/°C		
Change in temperature ΔT/°C		

Based on using $100\,cm^3$ of water with a specific heat capacity of $4.2\,J\,K^{-1}\,g^{-1}$

ⓔ Note here that we make the assumption that $1\,cm^3$ of water has a mass of $1\,g$.

Energy released = $100 \times 4.2 \times \Delta T$ ✓

Number of moles of fuel = $\Delta m/74.1$ ✓

ⓔ 74.1 is the mass of 1 mole of both fuels.

Enthalpy of combustion = energy released/number of moles of fuel

(b) Give *three* reasons why your experimental values are likely to be less negative than those shown in the table. (3 marks)

(b) ■ Heat lost to surroundings ✓

■ Incomplete combustion of fuel ✓

■ Water produced as a vapour in experiment and not a liquid, which is required for standard conditions ✓

Question 23

Ethanol can be prepared by the enzymatic fermentation of glucose using baker's yeast as a source of the necessary enzyme.

$C_6H_{12}O_6(aq) \rightarrow 2C_2H_5OH(aq) + 2CO_2(g)$

(a) You are provided with 5.0 g of glucose 1.0 g of baker's yeast and standard laboratory apparatus. Give details of how you would produce a *pure* sample of ethanol.

Your method should include:
■ a labelled diagram of your apparatus for the fermentation process
■ a clear step-by-step method
■ an indication of how the ethanol is separated and made as pure as possible (6 marks)

ⓔ You should be familiar with the basic idea behind enzymatic fermentation as represented in the equation above. In this case you are required to produce a pure sample of ethanol. Thus once the ethanol is collected by fermentation, there will need to be a further drying process in order to remove any remaining water. This is necessary as it is not possible to get ethanol of any greater purity than 96% *by fractional distillation alone*. Marks can be awarded for the diagram and/or for the method.

Student answer

(a)

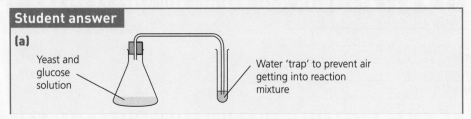

Yeast and glucose solution

Water 'trap' to prevent air getting into reaction mixture

e Your diagram should show the reaction flask containing glucose solution and yeast ✓ and then a water trap to prevent any air getting in to oxidise ethanol ✓ (must be anaerobic conditions).

> Temperature should be controlled to be the optimum for fermentation (e.g. between 25°C and 40°C). ✓
>
> It should be acknowledged that fermentataion takes at least 2 to 3 days ✓ and that it can be seen that fermentation stops when there are no more bubbles in the water trap ✓.
>
> Once fermentation has stopped, the reaction mixture can be filtered and then ethanol can be extracted by fractional distillation ✓.
>
> Distillate must be futher purified by adding a chemical drying agent (e.g. anhydrous calcium chloride or sodium sulfate) ✓.

e Any 6 marks out of 7.

(b) Calculate the atom economy for this reaction. (1 mark)

> **(b)** Mass of reactants, M_r glucose = 180. 2 mol of ethanol produced = 2 × 46 = 92.
>
> So atom economy = (92/180) × 100 = 51% ✓

(c) Assuming a 35% conversion, calculate the mass of ethanol collected. (2 marks)

> **(c)** 5 g of glucose = 5/180 = 0.0278 mol. Two moles of ethanol produced per mole of glucose, so 2 × 0.0278 = 0.056 mol ✓. Based on a 35% conversion (35/100) × 0.056 = 0.0194 mol of ethanol produced. M_r ethanol = 46, so 0.0194 × 46 = 0.89 g of ethanol ✓.

(d) Suggest why the percentage conversion might be as low as 35% in this synthesis. (2 marks)

> **(d)** Yeast is a living organism and ethanol is excreted as a by-product of anaerobic respiration ✓. Once the concentration of ethanol is sufficiently high, the high ethanol concentration will kill the yeast ✓.

A
acids, weak
    exam question 81–2
    K_a value 32–5
activation energy, of a reaction 53–6
Arrhenius equation 53, 54–5
aspirin synthesis 59–62

B
batteries
    see electrochemical cells
buffer solutions 32–5
    exam question 81–2

C
chlorination, 2-methylpropan-2-ol 23–6
'clock reactions'
    exam question 78–9
    rate equations 49–52
combustion, enthalpy of 92–3
command terms 63
complexes, transition metal 41–4
concentration of a solution
    hydrochloric acid 13–15
    sodium hydroxide 9–13
constants
    equilibrium 53
    rate 53
continuous monitoring method 45
covalent bonds 15

E
electrochemical cells 35–8
    exam question 79–80
enthalpy change of reaction
    exam question 67–9
    Hess's law 29–32
enthalpy of combustion 92–3
equilibrium constants 53
ethanol
    exam questions 66–7, 93–4
    oxidation 19–22

F
first law of thermodynamics 29

G
gases, molecular volume 5–9

H
halogenoalkanes, hydrolysis of 15–19
Hess's law
    enthalpy change of reaction 29–32
    exam question 67–9
H^+ ion (proton donor) 6
hydrochloric acid
    concentration of a solution 13–15
    exam question 87–8
hydrolysis
    exam question 65–6
    of halogenoalkanes 15–19

I
inorganic unknowns, analysis 26–9, 57–9
'iodine clock' reaction, exam question 78–9
iodine–propanone reaction, titrimetric method 44–8

K
K_a value, for a weak acid 32–5

L
laws of thermodynamics 29
ligands 41
lone pairs 44

M
metal complexes 41–4
2-methylpropan-2-ol, chlorination 23–6
molar volume, of a gas 5–9

N
neutralisation reactions 9–15

notation, electrochemical cells 35–6
nucleophiles 16

O
order of reaction 45
organic unknowns
    analysis 26–9, 57–9
    exam question 74–5, 84–5
oxidation, ethanol 19–22
oxidising agents 20

P
proton donor (H^+ ion) 6

Q
qualitative analysis 26
    exam question 70–2
quantitative analysis 26
    exam question 70–2

R
rate constants 53
rate equations
    'clock reactions' 49–52
    exam question 78–9
    titrimetric method 44–5
rates of reaction 44–8
reaction mechanisms 44
reactions, activation energy 53–6
redox titration 38–41
    exam question 85–6

S
salt bridges 37
    exam question 79–80
Science Practical Endorsement 63
sodium hydroxide
    concentration of a solution 9–13
    exam question 87–8
specific heat capacity 31
standard solution 9, 10
starch indicators 45
stoichiometric equations 55

Index

T

thermodynamics, first law of 29

titration 11, 13

    exam question 85–6

titrimetric analysis 38–41, 44–8

    exam question 64–5

    hydrochloric acid 13–15

    sodium hydroxide 9–13

transition metal complexes 41–4

V

volume, molar 5–9

W

weak acids, exam question 81–2